PRINTED FOR THE MALONE SOCIETY
BY DAVID STANFORD AT THE
UNIVERSITY PRINTING HOUSE
OXFORD

COLLECTIONS
VOLUME XIV

THE MALONE SOCIETY
1988

THIS edition of seven Jacobean academic plays was prepared by Suzanne Gossett and Thomas L. Berger and checked by H. R. Woudhuysen and the General Editor.

The Society is grateful to the Folger Shakespeare Library for permission to edit texts from its manuscripts, J.a.1 and J.a.2, and reproduce plates from the same.

November 1988 JOHN PITCHER

JACOBEAN ACADEMIC PLAYS

CONTENTS

PREFACE

MANUSCRIPTS J.a.1 and J.a.2 in the Folger Shakespeare Library, Washington DC, are early seventeenth-century collections of poetry, prose, and dramatic texts in Latin and English. The manuscripts are not included in de Ricci's catalogue of the Folger holdings. They were described in 1959 by R. H. Bowers (whose description is cited hereafter as 'Bowers', together with page number).[1] Some of the texts in J.a.1 and J.a.2 are apparently unique; others are known in a number of surviving copies. As collections, the manuscripts give a general picture of the intellectual interests of the university-educated Englishman in the reign of James I.

The Latin and English dramatic pieces, with the exception of an abbreviated version of Ben Jonson's *Christmas his Masque*, all seem to be written with an academic audience in mind. The plays vary considerably in scope. Some are fully developed and formal, for instance, *Periander*, a St John's College, Oxford verse tragedy in English, or *Melanthe*, a Latin pastoral acted before King James at Trinity College, Cambridge in 1615. Others are 'shews', less traditional in form, and more raucous in humour. Generally shorter, these too vary, from *A Christmas Messe*, which conceals a five-act structure beneath its horseplay, to the ninety-line dialogue between Preist the Barbar and Sweetball his Man, which could easily have been played after dinner in a college hall. This edition collects and makes available those plays in English which have not been edited in modern times.[2]

The manuscripts, and the plays, have considerable similarities, but each manuscript has a separate history, and is described below individually. The texts of the plays are accompanied by headnotes which supply historical, bibliographical, and other details.

Among the scholars who have assisted the editors in the preparation of this volume are Peter Beal, John R. Elliott Jr., K. J. Höltgen, Hilton Kelliher, Pierre Lefranc, Alan Nelson, Richard Proudfoot, Eric Sams, R. B. Weller, and Laetitia Yeandle. The editors have received support from Loyola University of Chicago, St Lawrence University, and the Newberry Center for Renaissance Studies. Thanks are due to the staff of the Folger, for their many kindnesses.

[1] 'Some Folger Academic Drama Manuscripts', *Studies in Bibliography*, xii (1959), 117–30.

[2] Since the editors began their research, David L. Russell has edited three of the plays, *Heteroclitanomalonomia*, *Gigantomachia*, and *A Christmas Messe*, for a doctoral dissertation ('Stuart Academic Drama: An Edition of Three University Plays') at Bowling Green University (1979). This edition has been published, without alterations and under the same title, by the Garland Press (New York and London, 1987). R. S. Thomson, arguing that *Boot and Spurre* is a 'Jacobean quête', has included an edited text of that play in an article in *English Literary Renaissance*, xviii (1988), 275–93.

EDITORIAL CONVENTIONS

THE following conventions are used for editing plays from the manuscripts J.a.1 and J.a.2. Square brackets enclose deletions, except for those around folio numbers ([FOL. 19a], etc.); angle brackets enclose material which other causes (paper damage, blotting) have removed or made impossible to decipher. In both categories, dots indicate illegible characters (thus ⟨.⟩). Long *s* has been lowered, and all forms of the ampersand appear as &. The symbols for *per*/*par*, *pro*, and *pre* are represented as ꝑ, ꝓ, and pͬ respectively: the symbols for *uer* and *ner* are rendered as uͬ and nͬ (for example, 'genͬall', *Heteroclitanomalonomia*, l. 714). As with any transcription, several letters require editorial judgements, especially in distinguishing between majuscule and minuscule forms. Ruled margins go unnoticed.

Line numbering is continuous in each text, including scene directions but excluding catchwords. Where the title of a play is written in the same hand as that in the body of the text, it is included in the line count; the title is not given a number where it has been added in another hand (as is the case with three plays in J.a.1). Partial lines, centred and marginal directions, horizontal lines, and indented material have been oriented in relation to the nearest preceding full line. In numbering lines, slight misalignments of stage directions have been ignored, and in only the clearest cases have they been given separate line numbers. The length of lines in several plays is so great that it has proved necessary to use turn lines. In these cases, lines are numbered thus (*Ruff, Band, and Cuff*; see p. 134 and Plate 6):

B. Lace mee, thou wouldst bee laced thy selfe, for this is the very truth ruff, thou
⟨ar⟩t a plaine Knaue.

FOLGER MANUSCRIPT J.a.1

I. PROVENANCE AND DATE

MANUSCRIPT J.a.1 is catalogued in the Folger as a 'Dramatic and poetical miscellany, ca. 1600–ca. 1620'. The manuscript is a compilation of a number of separate booklets written at different times in different hands. At some time, probably in the late seventeenth century, these booklets were wrapped in a few leaves of paper (see collation below) and bound. The leather binding was matched in size to the largest of the booklets. At present the manuscript is once again in several sections: the covers are off, and several booklets and some individual pages are detached.

The manuscript was acquired from Maggs Brothers in 1933. It appeared as item 204 in *Maggs Catalogue* 569 (1932) as 'Contemporary Manuscript Plays and Masques of the Time of Shakespeare . . . From the Library of the Marquess of Cholmondeley'.[1] Maggs, in turn, had bought the manuscript at a Sotheby's auction 2 years earlier.[2] Neither Maggs nor Sotheby's can supply any further information about the early history of the manuscript.

The book-plate is listed by Augustus W. Franks as number 5811, Cholmondeley Library (printed on rose-coloured paper) Pictorial Armorial.[3] It depicts the front and side of a castle, with 'Cholmondeley Library' inscribed below the crenellations. Two shields on the front show two sheaves over a helmet, a variation of the Cholmondeley family crest, which shows two helmets over a sheaf. The only Cholmondeley book collection which has been given much attention is that of Reginald Cholmondeley of Condover Hall, Salop, a collection that was sold in 1887.[4] His book-plate reveals another variation of the

[1] *English Verse and Dramatic Poetry from Chaucer to Burns* (London, 1932), pp. 58–61.

[2] *Valuable Printed Books and Manuscripts . . . Comprising the Property of the Marquess of Cholmondeley* (London, 17–20 March 1930), item 450, p. 54.

[3] E. R. J. Gambier Howe, *Catalogue of British and American Book Plates Bequeathed . . . by Sir Augustus Wollaston Franks* i (London, 1903), 205.

[4] This collection was sold with 'other papers at Puttick and Simpson's 24 Nov 1887 lots 975 and 976 and bought by Quaritch. Two thousand documents [from this] relating to the Smyth family of Nibley . . . were sold on immediately to F. A. Crisp.' (*Guide to the Location of Collections described in the Reports and Calendars Series 1870–1980*, Historical Manuscripts Commission (London, 1982), p. 13). It may be of importance that in the collection of John Smyth of Nibley, the antiquarian (d. 1640), there was at least one manuscript play, a nine-scene Latin piece called *Sanctus Tewdricus* (Fifth Report of the Royal Commission on Historical Manuscripts, Part 1 (London, 1876), p. 340). Much of this information comes from Peter Beal. See also W. C. Hazlitt (ed.), *An Alphabetical Roll of Book Collectors from 1316–1898*, in Bernard Quaritch, ed., *Contributions towards a Dictionary of English Book Collectors*, 14 parts (1892–1921) in one vol. (London, 1921), Part 12 (1898), p. 8.

family shield. The rose-coloured library book-plate apparently dates from the nineteenth century and was probably that of the second Marquess, George Horatio (1792–1870). The 1932 sale occurred during the time of the fifth Marquess, George Horatio Charles (1883–1968), who had become Marquess in 1923. It is impossible to determine from the information available how long the manuscript had belonged to the Cholmondeleys.

Since the plays included in J.a.1 appear to have been intended for university performance, one might hope that enough evidence existed (as it does for J.a.2) to associate the manuscript definitively with either Oxford or Cambridge. This is not the case: J.a.1 includes enough material connected with the court as well as both universities to make any single provenance for the manuscript most unlikely.[1] Among the dramatic material in the manuscript, *Periander* is an Oxford play, and it is suggested below that *Risus Anglicanus* may well be connected with King's College, Cambridge. As for the four J.a.1 plays in this edition, none of them can be irrefutably linked to one of the universities. The most persuasive connection is the similarity between *Boot and Spurre* and the J.a.2 plays, *Ruff, Band, and Cuff* and *Gowne, Hood, and Capp*. Hilton Kelliher suggests that all three of these may be connected to the tract, *Worke for Cutlers. Or, a Merry Dialogue betweene Sword, Rapier, and Dagger. Acted in a Shew in the famous Vniuersitie of Cambridge* (London, 1616), and he further suggests that they may 'emanate from the same institution—possibly even from the pen of a single writer'.[2]

Scholars have traditionally ascribed *Boot and Spurre*, *Heteroclitanomalonomia*, *Gigantomachia*, and *A Christmas Messe* to Cambridge. The editor of the forthcoming Oxford volume of *Records of Early English Drama*, John R. Elliott, Jr., finds no external evidence that these plays were produced at Oxford, nor specific similarities between these plays and Oxford dramatic performances.[3] Furthermore, certain allusions (to 'Heteroclits' in *Gigantomachia*, and to a 'Gygantomachy' in *A Christmas Messe*) *may* indicate that *Heteroclitanomalonomia*, *Gigantomachia*, and *A Christmas Messe* were all produced in the same college (see below, p. 98). In all, and with due regard to the paucity of evidence, it seems reasonable to concur, tentatively, in assigning these four J.a.1 plays to Cambridge.

It is still important to emphasize that the evidence is meagre and largely

[1] Members of the Cholmondeley family who went to university went to Oxford. Indeed, Robert Cholmondeley, the first Viscount, was created a Baronet in 1611, the year he matriculated at The Queen's College, Oxford. Hugh Cholmondeley, the third Viscount, was at Christ Church (matriculated 1678), as were the fourth Viscount (second Earl, matriculated 1680) and the third Marquess (matriculated 1818).

[2] Private communication to the editors.

[3] Ibid.

inconclusive. G. E. Bentley perhaps displays too little caution when he writes of *A Christmas Messe* that

> the tone of the piece, its length, and the general device of impersonating items at a banquet are so much like Thomas Randolph's *Salting* . . . that the occasion of its production must have been a similar college banquet and a Cambridge college is therefore suggested.[1]

On this basis Bentley suggests a Cambridge origin for *Gigantomachia* and a shared 'informal, intimate character' between Thomas Randolph's *Salting* and *Boot and Spurre*.[2] Bentley has been followed cautiously by Schoenbaum in the revised edition of Harbage's *Annals*, which gives 'Cambridge(?)' as the origin of *Gigantomachia*, *Heteroclitanomalonomia*, and *A Christmas Messe*.[3]

Recent study of the form of two saltings (Randolph's and another from St John's College, Cambridge) gives us reason to think that *A Christmas Messe* is of a quite different genre. 'Saltings', we are told, 'were jocular ceremonies for the initiation of Freshmen students . . . [in which] a senior student played the role of the Father, while a select group of Freshmen played his Sons.'[4] In Randolph's salting the Father satirizes each Freshman while burlesquing 'the public exercises and subscriptions to the various oaths required for admission to the university'.[5] In this piece, each student becomes some element of the banquet, while in the St John's salting each Freshman becomes a part of the body. A salting has no plot. By contrast, *A Christmas Messe* has an absurdly complete, although miniaturized, five-act structure, and a plot very like the one in *Heteroclitanomalonomia*. There the nouns and verbs go to war for priority, whereas here Beef and Brawn claim precedence at the Christmas Feast, enlist troops, fight, and are finally reconciled. *A Christmas Messe* is one more example of the prevalence of the debate form in the university drama of the period.

The Cambridge identification may be questioned on other grounds. Richek points out that ceremonies similar to saltings were held at Oxford as well as Cambridge.[6] In addition, David L. Russell has claimed that two words used in *A Christmas Messe*, 'fly' and 'sconce', were Oxford words.[7] This is an intriguing

[1] G. E. Bentley, *The Jacobean and Caroline Stage*, 7 vols. (Oxford, 1941–68), v. 1306.
[2] Ibid. v. 1343 and 1295.
[3] Alfred Harbage, *Annals of English Drama 975–1700*, revised S. Schoenbaum (London, 1964), pp. 103 and 111.
[4] Alan H. Nelson, Appendix 12, 'Saltings', *Cambridge* (REED: forthcoming).
[5] Roslyn Richek, 'Thomas Randolph's *Salting* (1627), Its Text, and John Milton's Sixth Prolusion as Another Salting', *English Literary Renaissance*, xii (1982), 103–31 (esp. 105). See also Alan H. Nelson, 'A Salting at St. John's', *The Eagle* [of St John's College, Cambridge], lxix (1983), 23–30.
[6] Richek, p. 105.
[7] David L. Russell (ed.), *Stuart Academic Drama: An Edition of Three University Plays* (New York and London, 1987), p. 37.

suggestion, but not a decisive one. A 'sconce', which was the 'fine of a tankard of ale . . . imposed by undergraduates . . . for some breach of customary rule', or a fine imposed by university officials, is identified in the *OED* as 'Oxford (?formerly also at Cambridge)'. And the meaning of Vinegar's threat, 'Beeware King Brawne Ile kill thee, with a fly' (l. 200), is ambiguous. Russell ingeniously suggests that 'fly' here refers to the 'festival formerly observed by the Oxford cooks' (*OED* 8). This is certainly appropriate to the context, but the word could mean several other, quite different, things (for example, 'fly' *OED* 5 is defined as 'a familiar demon'). Nevertheless, the conjunction of the two words is worth weighing against the possibility that *A Christmas Messe* is a Cambridge play.

The manuscript cannot be dated precisely. The earliest material in it is Edmund Campion's funeral oration for Sir Thomas White, who died in 1566/7; the copy in J.a.1 could have been made at any time in the next half century. The latest piece is almost certainly the first one. Andrew Melville's *Pro Supplici*, or *Anti-Tami-Cami-Categoria* was written in 1603/4 but not published until 1620 (STC 4361). The epigrams of reply, *Musae Responsoriae* (the second item in the manuscript), are by George Herbert. His editor believes that most of this material was composed in response to the printed (1620) version of Melville's work, and that a *terminus ad quem* is furnished by Melville's death in 1622.[1] No piece in the manuscript is later than this, and a number of works—for example, the *Secundum Iter Boreale* and *A Christmas Messe*, both from 1619—are grouped around this date.

II. CONTENTS

The description of J.a.1 offered by Bowers in 1959 was incomplete: further research has enabled the editors to correct some of his identifications and to propose new ones. Corrections have also been made to Bowers's bibliographical information, but his numbering of J.a.1's contents has been retained. The separate pieces are as follows:

1. *Pro Supplici Evangelicorum Ministrorum in Anglia . . . sive Anti-Tami-Cami-Categoria*. Fols. 2a–5a. [1604]

   The author of this piece is Andrew Melville, Principal of New College, St Andrews. The first printing is in *Parasynagma Perthense et Ivramentvm Ecclesiæ Scoticanæ*, 1620 (STC 4361). It also appears in David Calderwood's *Altare Damascenum* (Amsterdam, 1623; STC 4353).

2. *Musae Responsoriae*. Fols. 7a–17b. [1620–2]

   These Latin epigrams are by George Herbert, who may have begun them

[1] F. E. Hutchinson (ed.), *The Works of George Herbert* (Oxford, 1941), pp. 587–8.

while he was a student at Westminster School. It appears that they were largely written between 1620 and 1622, that is, between the publication of Melville's work and his death.[1] The J.a.1 text does not have the epigram addressed to Lancelot Andrewes, Bishop of Winchester, which, in the version published in 1662 in *Ecclesiastes Solomonis* by James Duport (Wing STC V669), appears between the dedication to Prince Charles and the first epigram.[2] Andrewes became bishop of Winchester in February 1618/19.

3. *Boot and Spurre*. Fols. 19a–23a.

Discussed below, pp. 21–2.

4. *Risus Anglicanus*. Fols. 24a–43b. [1614–18]

Bowers (p. 118) writes that this 'anonymous Latin comedy . . . related the alleged tribulations and failures at Rome of the leading Jesuit publicists to counteract the, to them, dangerous implications of the Oath of Allegiance of 1606'. *Risus Anglicanus* is clearly an academic play, using the system of houses which prevailed at the universities: 'Inscriptiones in Scaena: Ignatianum, Forum Romanum, Vaticanum, Forum exoticum, Ianua Ditis.' The majority of the speakers are historical participants in the controversy, although Daemunculi Assistentes and Lucifer also appear. The striking conclusion of the play brings on the English Jesuit Thomas Fitzherbert, who speaks English because he is too ignorant to speak Latin.

Risus Anglicanus may well be connected with Samuel Collins (1576–1651), DD Cambridge 1612/13, Provost of King's College, Cambridge, 1615, and Regius Professor of Divinity, 1617. For a decade Collins was involved in a quarrel very much like the one depicted in the play. The relevant sequence of events is as follows: Cardinal Bellarmine (under the pseudonym of Matthaeus Tortus) and Eudaemon-Johannes had attacked Bishop Lancelot Andrewes, who was engaged in defending the Oath and the King's books about it. Collins in 1612 counter-attacked these two in his *Increpatio Andreæ Eudæmono-Johannis Jesuitæ* (STC 5563). In 1613 Thomas Fitzherbert, for many years the Clergy Agent in Rome, attacked Andrewes's answer to Bellarmine, and to this Collins in turn replied in his *Epphata to F.T. Or, the defence of the lord Bishop of Elie . . . concerning his answer to Cardinall Bellarmines Apologie*, 1617 (STC 5561). To this Fitzherbert replied yet again, in 1621, in *The Obmutesce of F.T. to the Epphata of D. Collins* (STC 11020).[3]

The author of the play is well informed about the controversy, putting on stage most of the major opponents of the Oath: Bellarmine as Tortus,

[1] Ibid. 587–9.

[2] Ibid. 384–5.

[3] The whole affair is described in detail by C. H. McIlwain in his edition, *The Political Works of James I* (Cambridge, Mass., 1918), pp. xlix–lxxx.

Eudaemon-Johannes, Suarez, and Martin Becan. It is, however, the focus on Fitzherbert, who was actually a minor figure internationally, which suggests the link with Collins. According to the *DNB*, Collins was 'reckoned the most fluent Latinist of his age, and was remarkable for his admirable wit and memory'. It is possible that Collins wrote *Risus Anglicanus* himself; at the least, it would have been well received in his honour at King's. It is hardly a coincidence that Collins was made Regius Professor in 1617, the year which saw the appearance of his *Epphata*.

Risus Anglicanus must date after 1614, the year when Suarez's book was burned in Paris (v. ii). The Fitzherbert references suggest that the play was written before 1618, when Fitzherbert became Rector of the English College in Rome, a post he held until his death in 1639. The play makes no mention of this position. Since Fitzherbert is specifically mocked for ignorance, the author would have been unlikely to miss the opportunity, had it existed at the time of writing, to extend the attack to the College (a favourite target of English anti-Catholic writers).

5. *Funebris Oratio in Obitum Thomae White* . . . per Edmundum Campianum. Fols. 44a–46a.

White died 12 February 1566/7, and Campion was executed in 1581.

6. *Oneirologia*. Fols. 47a–63a. Dated 20 November 1605.

This 'breife discourse of the nature of Dreames' is by Richard Haydocke of New College, Oxford, who gained notoreity by preaching in his sleep. King James himself investigated the case and was reported to have uncovered the fraud by shouting 'God's wounds, I will cutt off his head', at which Haydocke 'leaped in terror from his couch'.[1] Haydocke was brought to court on 22 April 1605 and confessed 6 days later. The treatise begins with a letter to the King, comparing James to Solomon and begging pardon for Haydocke's error. The letter which follows, 'To the Curteous Reader', implies that Haydocke expected to publish the work, partly to dispel uninformed 'sinister censures'. In the treatise Haydocke admits that he had 'perfect knowledge' of what he was saying, and he takes up the general question of 'how farre the soule exerciseth her operations in time of sleep'. He also 'presumes' to add the King's 'most forcible Arguments . . . that there can be noe reasonable discourse in Sleepe'.

Weller and K. J. Höltgen agree that the manuscript is in 'Haydocke's neat italic hand'.[2] Haydocke also transcribes item 7 below ('*Reasons . . . to induce*'), perhaps, as Höltgen suggests, 'to prove himself a good and loyal subject of the

[1] D. Harris Willson, *King James VI and I* (London, 1956), p. 310.

[2] Ralph B. Weller, 'Some Aspects of the Life of Richard Haydocke, Physician, Engraver, Painter, and Translator (1569–?1642)', *The Hatcher Review*, ii (1985), 464.

King . . . after the suspicions he had aroused through his activity as a nocturnal preacher'.[1]

7. *Reasons . . . to induce . . . this Kingdome to graunt unto the King . . . a large subsidie*. Fols. 71a–80b. Dated 1617, attributed to Dr Willet.

Andrew Willet, who was Chaplain in Ordinary and Tutor to Prince Henry, was also interested in anti-Catholic controversy. In 1592 he had written *Synopsis Papismi* (STC 25696, with five editions, all augmented, to 1634), a reply to Bellarmine. In *Reasons . . . to induce* he again raises the issue behind *Risus Anglicanus*, that is, 'the Spanish Doctors holdinge, yt it is lawfull for any man to kill his Kinge denounced an Heretick by ye Pope', and he appends the marginal note 'Simanca Mariana Suarez' (Fol. 79a).

8. *Iter Boreale*. Fols. 83a–91b.

This Latin poem was written by Richard Eedes, *c.*1584. Eedes was BA Oxford, 1574, and later Dean of Worcester.

9. *Secundum Iter Boreale*. Fols. 93a–99b.

This English poem is by Richard Corbett (1582–1635). Although the journey ends at St John's College, Oxford, Corbett himself lived at Christ Church from 1598 to 1628. Bennett and Trevor-Roper suggest that the poem was probably written in 1619, before Corbett's promotion to Dean in 1620, but Peter Beal regards the date as 'controversial' because Brian Duppa appears in certain manuscript versions of the poem.[2] Among the manuscripts cited by Beal, CoR 313 actually dates the 'Itinerarij' as '10 Aug. 1618'.[3]

10. *A Christmas Messe*. Fols. 105a–115b. Dated 1619.

Discussed below, pp. 34–5.

11. *Heteroclitanomalonomia*. Fols. 119a–133a. Dated 1613.

Discussed below, pp. 57–60.

12. *Periander*. Fols. 134a–157b. Described as 'mr John Sansburyes'.

Periander was the last play acted (13 February 1608) in the Christmas revels at St John's College, Oxford, 1607–8. Alfred Harbage discusses the relationship between the Folger (J.a.1) text of *Periander* and the St John's version printed by Frederick Boas and W. W. Greg in *The Christmas Prince* (Malone Society, 1923).[4] Working from the text supplied by Boas, Harbage concludes that the Folger manuscript is based on the St John's one because it silently incorporates a marginal emendation. But the Folger manuscript also

[1] Private communication to the editors, 3 August 1988.

[2] J. A. W. Bennett and H. R. Trevor-Roper (eds.), *The Poems of Richard Corbett* (Oxford, 1955), p. 118; Peter Beal, private communication to the editors, 4 June 1987.

[3] See Peter Beal, *Index of English Literary Manuscripts*, vol. 2, pt. 1 (London, 1987), p. 181. Manuscripts of the poem are listed on pp. 179–81.

[4] 'The Authorship of *The Christmas Prince*', *Modern Language Notes*, l (1935), 501–5.

includes a different final line and it has many small variations: it is possible that both manuscripts derive from a common source.

In the right-hand margin of the title-page of the Folger manuscript (Fol. 134a) there is an annotation which Harbage reads as 'Englishs Fairre copy'. The present editors read this as 'Englishs Joannasis', which still fits Harbage's suggestion that the reference is to John English, 'who was an undergraduate of the College, and the "Lrd Cheife-Iustice Examener of all causes capitall" in the household of the [Christmas] Prince'.[1]

13. *Especiall Notes concerning her Majesties Navie*. Fols. 161a–165b.

This is an apparently unique Elizabethan version of Ralegh's *Observations Concerning the Royal Navy and Sea-Service*, also known as *Notes on the Navy* and previously assumed to date from after 1608. The Elizabethan version was written in late 1597 or early 1598, after Ralegh's return to court following the Islands Voyage. The manuscript differs from the Jacobean version by including a preface to the Queen and a conclusion asking that she take action to remedy the problems of the navy which the author has identified. In the Jacobean version the conclusion justifies 'keeping and maintaining so great a navy . . . the times being now peaceable'. The later version also incorporates Ralegh's 1608 letter to Prince Henry about the construction of a warship.[2]

Neither Lefranc nor Beal has found any other pre-1603 version of the *Notes*.[3] Just how the Folger manuscript came to preserve the Elizabethan version is unclear. Although the presence of item 15 below (*A Discourse*) suggests the compiler's interest in Ralegh's work, he may not have known the identity of the author of the *Notes*, since no name is given.

14. *Christmas his Showe*. Fols. 168a–174b. Dated 1615.

This is an abridged version of Ben Jonson's *Christmas his Masque*, which was performed at court 1616/17. In the full version Christmas asserts that the masque was originally intended for 'Curryers Hall' but that he 'thought it convenient, with some little alterations, and the Groome of the *Revells* hand to't, to fit it for a higher place'.[4] This may reveal that the masque, which is actually a rather simple mummers play, was written originally in 1615 in circumstances now lost. The manuscript annotation 'The christmas shewe

[1] 'The Authorship of *The Christmas Prince*', *Modern Language Notes*, l (1935), 504.

[2] For a full discussion of the differences, see Suzanne Gossett, 'A New History for Ralegh's *Notes on the Navy*', *Modern Philology*, lxxxv (1987), 12–26. Pierre Lefranc (correspondence with Suzanne Gossett, 29 February 1984) agrees with the dating.

[3] Pierre Lefranc, *Sir Walter Ralegh Écrivain* (Paris, 1968), pp. 53, 58–9; Peter Beal, *Index of English Literary Manuscripts*, vol. 1, pt. 2 (London, 1980), pp. 433–4.

[4] C. H. Herford, Percy and Evelyn Simpson (eds.), *Ben Jonson*, 11 vols. (Oxford, 1925–52), vii. 438 (Jonson's full text of the masque is on pp. 437–47.)

before the kinge. 1615' indicates a court performance, however, and '1615' may be merely a slip of the pen, or other error.

15. *A Discourse touching a Marriage betweene Prince Henry of England and a daughter of Savoye*. Fols. 175a–182b. Dated 1611.

This tract is by Sir Walter Ralegh. In spite of the manuscript date, Lefranc suggests that it was written 'entre le début de l'été et le 6 novembre 1612'.[1]

16a. *Convivium Philosophicum*. Fol. 183a–b. [1611]

The authorship of this Latin poem has not been established. Osborn attributes it to John Hoskyns (of New College, Oxford, 1585–95),[2] but Hoskyns appears in both the Latin and the English versions under a different pseudonym from that of the author, Rodolpho Colfabio (see Bowers, p. 123, and his references).

16b. Untitled English Version of item 16a, the *Convivium Philosophicum*. Fol. 184a–b.

The poem is attributed by Andrew Clark to John Reynolds, Fellow of New College, Oxford in 1600.[3]

17. *Vrbium Italicarum Descriptio Thomas Edwardi Angli*. Fol. 185a–b.

This series of Latin hexameters on the cities of Italy appeared in the *Parvum Theatrum Urbium* of Andriano Romano (Frankfurt, 1595). The major differences between the manuscript and the printed version are in the comments on the cities of Viterbo, Fano, and Vicentia, and the addition of four lines in the manuscript:

Laurentus properant Magnates quolibet anno
Fundentes lachrymas, munera, vota, preces.
Laurentus nudis pedibus plebs crebra frequentat
Quam mouet internus relligionis amor.

The comment in the margin is 'A city of Latiũ not far frõ Lauiniũ. Laurentus.' The Latin verses were reprinted and translated in Robert Vilvain's *Enchiridium Epigrammatum* (London, 1654; Wing STC V395). The last two lines in Vilvain are not present in the Folger text, nor in Romano. Thomas Edwards also wrote *Cephalus & Procris* and *Narcissus* (1595, STC 7525); the *DNB* finds 'some reason to suppose that [he] was an Oxford man'.

18. *Gigantomachia, or Worke for Jupiter*. Fols. 186a–200a.

Discussed below, pp. 98–100.

1 Lefranc, p. 35.
2 Louise B. Osborn, *John Hoskyns 1566–1638*, Yale Studies in English 87 (New Haven, 1937), p. 9.
3 Andrew Clark (ed.), *Aubrey's 'Brief Lives'*, 2 vols. (Oxford, 1898), ii. 53.

III. PAPER, HANDS, AND ANNOTATIONS

J.a.1 is a compilation of seventeen originally separate manuscripts. In its present state, the manuscript has thirty-two gatherings and sixteen watermarks. Most of the watermarks are very common ones (pots), and the identifications offered below are with the closest ones known. The table below shows how the manuscript is made up (retaining Bowers's numbering of items once more):

Item number	*Folio*	*Gathering*	*Watermarks*
	0.1–0.3	flyleaves and endpaper	none
1, 2	1–18	1 and 1a[1]	A: closest to Churchill 314, horn, but cf. Heawood 1216, crozier[2]
			B: Fol. 17 only, Heawood pillar 3500
3	19–23	2 (23 has no conjugate leaf)	C: pot, like Heawood 3561, 3576
4	24–43[3]	3	D: pot, like Heawood 3575 and 3582–3
5	44–46.1[3]	4	E: appears only on Fol. 44, unidentified bottom of pot
6 (and blank sheets)	47–50	5	F: pillar, like Heawood 3498
	51–54	6	F
	55–58	7	F
	59–62	8	F
	63–66	9	F
	67–70	10	F
7	71–82	11	G: crozier, Heawood 1226
8	82.1–85	12	H: pot, Heawood 3581
	86–89	13	H
	90–92.1	14	H
9	93–104	15	I: pot, Heawood 3584

1 Three conjugate leaves were inserted between Fols. 15 and 18.1, now a stub. The gathering collates: Fols. 1–18.1, 2–15, 3–14, 4–13, 5–12, 6–11, 7–10, 8–9, 15.1–16, 16.1–17, 17.1–18.

2 William Algernon Churchill, *Watermarks in Paper* (Amsterdam, 1935); Edward Heawood, *Watermarks Mainly of the 17th and 18th Centuries* (Hilversum, 1950).

3 Fols. 24–43 and 44–6 were once bound together independently of J.a.1, but stubmarks show that 24–43 were originally separate.

Item number	Folio	Gathering	Watermarks
10	105–118	16	J: Folger collection of watermarks, twisted pillars
11	119–126	17	K: pillar, Heawood 3499
	126.1–133	18	K
12	134–160.1	19	L: pot, close to Heawood 3647 with initials RG
13	160.2–167	20	M: pot, Heawood 3678
14	168–170.1	21	none
	171–172	22	N: pot, Heawood 3581–3582
	173–174	23	N
15	175–182	24	O: cockatrice: somewhat like Heawood 838, but smaller. This watermark is almost invisible.
16a, 16b	183–184	25	none
17	185	26	none
18	186–187	27	P: pot, Heawood 3562, 3565
	188–189	28	none
	190–193	29	P
	194–197	30	P
	198–199	31	P
18 ends	200–200.1	32	none
	200.2–200.4	flyleaves and endpaper	none

There are sixteen (possibly seventeen) main hands in J.a.1, as well as two kinds of annotators: those who annotated various pieces separately, and those who annotated the volume, or part of the volume, as we now have it. The main hands appear as follows:

Item number	Title	Folio	Hand
1, 2	*Pro Supplici*; *Musae*	2a–17b	A
3	*Boot and Spurre*	19a–23a	B (see Plate 1)
4	*Risus Anglicanus*	24a–43b	C
5	*Funebris Oratio*	44a–46a	D
6, 7	*Oneirologia*; *Reasons*	47a–80b	E (Haydocke)

Item number	*Title*	*Folio*	*Hand*
8	*Iter Boreale*	83a–91b	F
9	*Secundum Iter*	93a–99b	G
10	*A Christmas Messe*	105a–115b	H (see Plate 2)
11	*Heteroclitanomalonomia*	119a–133a	I (see Plate 3)
12	*Periander*	134a–157b	J
13	*Especiall Notes*	161a–165b	K
14	*Christmas his Showe*	168a–174b	L
15	*A Discourse*	175a–182b	M
16a	*Convivium Philosophicum*	183a–b	N
16b	English *Convivium*	184a–b	O
17	*Vrbium Italicarum*	185a–b	F′
18	*Gigantomachia*	186a–200a	P (see Plate 4)

Several pieces were annotated separately. *Risus Anglicanus* has annotations in its main hand C, which returns in another ink to gloss the manuscript. Many of these glosses take the form of making an *x* in the manuscript and then adding a Latin stage direction to correspond. On 20a a word is corrected in the text of *Boot and Spurre* by another hand. *Periander* is annotated in a mixed secretary and italic hand, Z, on 149b (speech prefix 'Reso') and on 151b, 152a, 152b, and 156b (marginal notes). On 153a J writes only the word 'Chorus' and Z writes the dialogue between Resolution and Detraction which completes the act. On 157b the Chorus by Resolution and Detraction is again written by Z after ten lines by J which conclude with flourishes. On 157b J writes the Epilogus. Beneath the flourishes which follow the last line, 'By many hands he may reviue agayne', Z writes 'your gentler hands will giue hym life agayne'. This is the last line in the St John's manuscript of *Periander*; it is possible that Z was comparing it with that version.

On 185b, in the Latin hexameters on Italian towns, an apparently different hand supplies the English names of the towns. Finally, the penultimate leaf of *Gigantomachia*, Fol. 199, has pasted on to its recto a cancel which changes the ending so that the giants are killed rather than stunned by thunder (see below, p. 120). Unlike Bowers, the present editors take this cancel to be in a hand different from that of the manuscript (see Plate 4).

Many hands have been at work in several sections of the manuscript. First of all, there are modern pencil markings. Inside the front cover, in what is probably a note by Maggs, is written 'From Library of Marquis Cholmondeley'. Also in pencil are 1b 'Andrew Melville, c. 1604'; 6b 'Andrew Melville'; 18b 'Boot and

Spurr'. These are probably recent. Other markings include: 46b a small piece of a design: 61a a design above the title; 66a a head in pencil; 100a a head in pencil, with this written above in ink over pencil: 'Paucae Antiquae Picturae & Michael Angelo Pictae'; 101a a head; 101b the words 'Pease cod from whence proceed', with, below, a sketch of a head facing a lectern holding a book; 103b a horse and keeper; 158b, 159a, 160a, 166b sketches of heads.

On the first folio there are ten lines written in shorthand (see Plate 5). Eric Sams believes the system to be that of Thomas Shelton, whose *Short-Writing* or *Tachy graphy* appeared in numerous editions from 1630 onwards. The author of this shorthand was evidently inspired by poems like the *Iter Boreale*. Dr Sams interprets the final line as 'We drink stale beer I think 'twas never new.'[1] The rest has not been deciphered.

Two early hands (X and Y), possibly mid-seventeenth-century ones, using ink which has faded to the same brown as most of the ink in the manuscript, have gone through the manuscript adding titles and other information. These annotations could have been made either before or after the manuscripts were bound. It is notable that the hands do not overlap, one appearing in the first half of the manuscript and the other in the second half. The first of these hands, X, adds to:

19a 'Boot & Spurre:/&c' (see Plate 1)
71a (to title) 'An. 1617, per Dtem willet./'
83a ':per Richardus Eeds./'
105a '1619'
119a '1613.' (This follows the black-letter title, 'Heteroclitanomalonomia', which could be in any hand.)

The second of these hands, Y, adds to:

134a '⟨J.P.P.⟩ Periander made bye mr John Sansburyes.'
134b A complete but unheaded *dramatis personae*, beginning 'Periander—the Kinge'
161a 'A tract concerninge the shippinge of England/'
168a 'Mock-maske The christmas shewe before the kinge. 1615.'
186a 'Gigantomachia: or'

Finally, in the right-hand margin of 134a appears the mysterious annotation 'Englishs Joannasis'.

[1] Private communication to the editors. The edition of 1630 appears to be the earliest extant text (STC 22404), although the Stationers' Register reveals that the book was entered on 17 April 1626.

It is worth noting that the early annotators either could not, or would not, identify the authors of the four plays in the present edition. The dates given for *A Christmas Messe* and *Heteroclitanomalonomia* suggest that the annotator, X, may have been aware of specific performances which he had either attended or heard about. It is reasonable to suppose that all of the plays included in the manuscript were performed, although much else about them must remain uncertain.

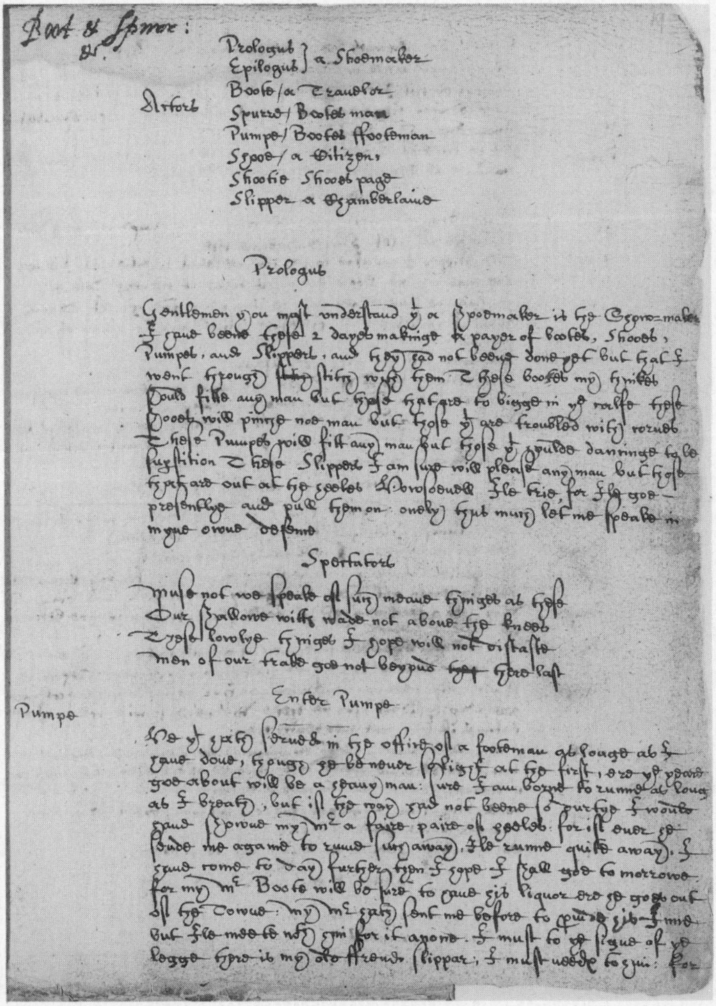

Boot & Spurre:

Prologus } a Shoemaker
Epilogus }
Boote / a Traveller
Actors
Spurre / Bootes man
Pumpe / Bootes ffooteman
Shoe / a Citizen
Shootie Shooes page
Slipper a Chamberlaine

Prologus

Spectators

Enter Pumpe

Pumpe

PLATE 1: FOLGER MANUSCRIPT J.a.1, FOL. 19a (TWO-THIRDS FULL SIZE): *BOOT AND SPURRE*, LINES 1 TO 38

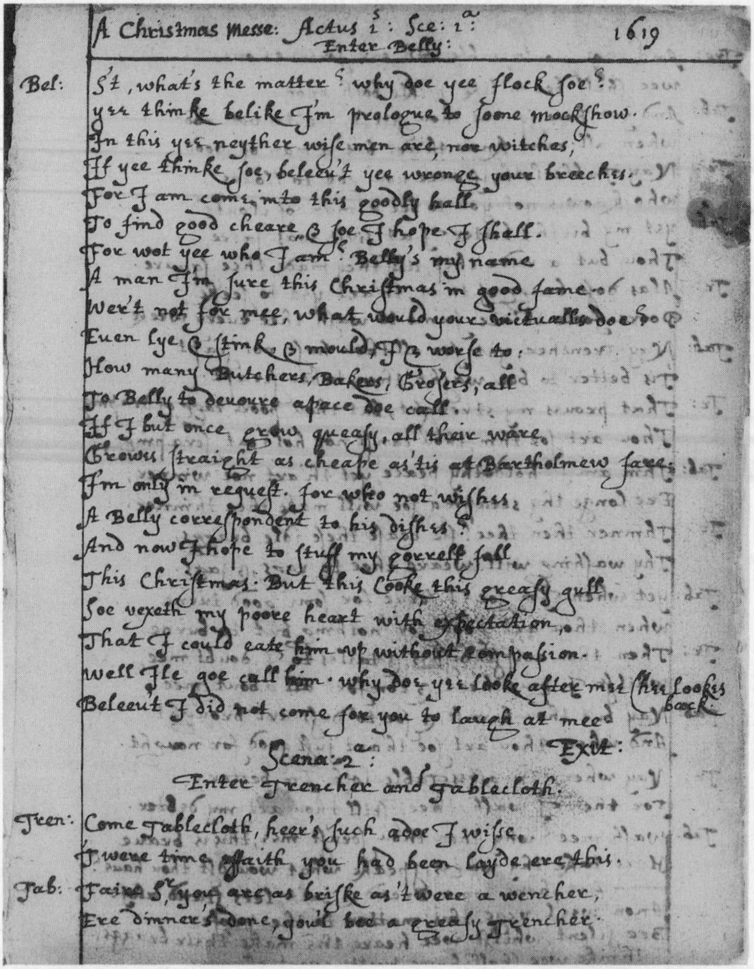
A Christmas messe: Actus 1s: Sce: 1a: 1619
Enter Belly:

Bel: Sr, what's the matter? why doe yee flock soe?
yee thinke belike I'm prologue to some mockshow.
In this yee neyther wise men are, nor witches;
If yee thinke soe, beleev't yee wronge your breeches.
For I am come into this goodly hall
To find good cheare & soe I hope I shall.
For wot yee who I am? Belly's my name
A man I'm sure this Christmas in good fame.
Wer't not for mee, what would your victualls doe?
Even lye & stink, & mould, & worse to:
How many Butchers, Bakers, Grosers, all
To Belly to devoure apace doe call.
If I but once grow queasy, all their ware
Growes straight as cheape as'tis at Bartholmew fare.
I'm only in request. for who not wishes
A Belly correspondent to his dishes?
And now I hope to stuff my gorrell full
This Christmas: But this Cooke this greasy gull
Soe vexeth my poore heart with expectation,
That I could eate him up without compassion.
Well Ile goe call him. why doe yee looke after mee? (he lookis back)
Beleev't I did not come for you to laugh at mee
Exit:

Scena 2a:
Enter Trencher and Tablecloth:

Tren: Come Tablecloth, heer's such adoe I wisse.
Twere time i'faith you had been layde ere this.
Tab: Faire Sr, you are as briske as 'twere a wencher,
Ere dinner's done, you'l bee a greasy Trencher.

PLATE 2: FOL. 105a (TWO-THIRDS FULL SIZE): *A CHRISTMAS MESSE*, LINES 1 TO 31

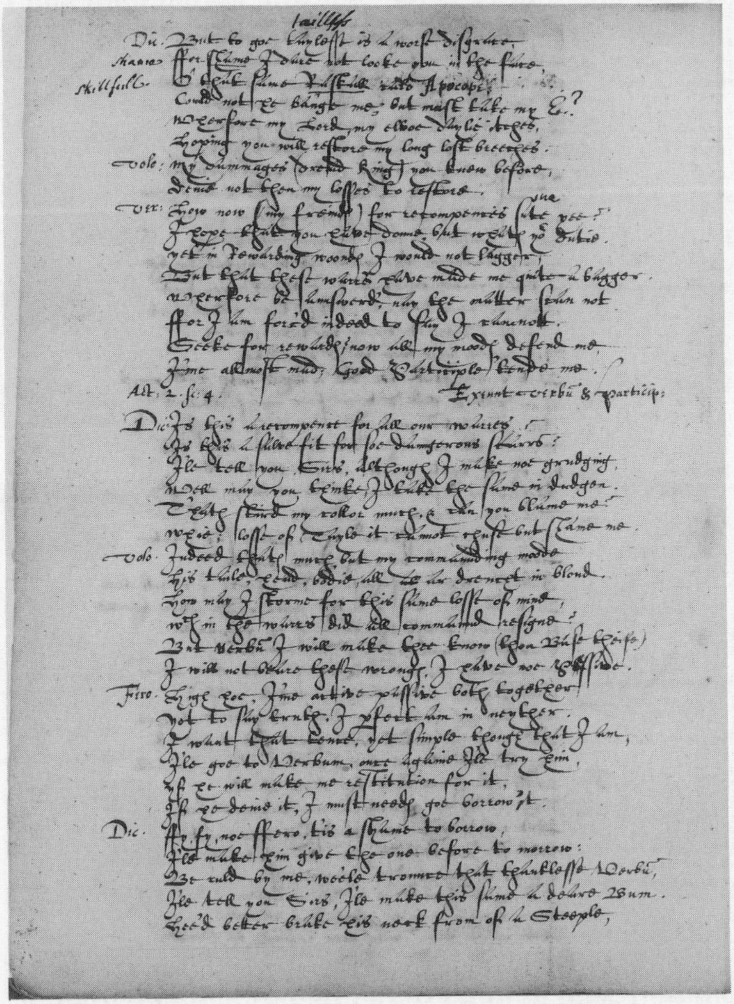

PLATE 3: FOL. 122b (TWO-THIRDS FULL SIZE): *HETEROCLITANOMALONOMIA*, LINES 279 TO 318

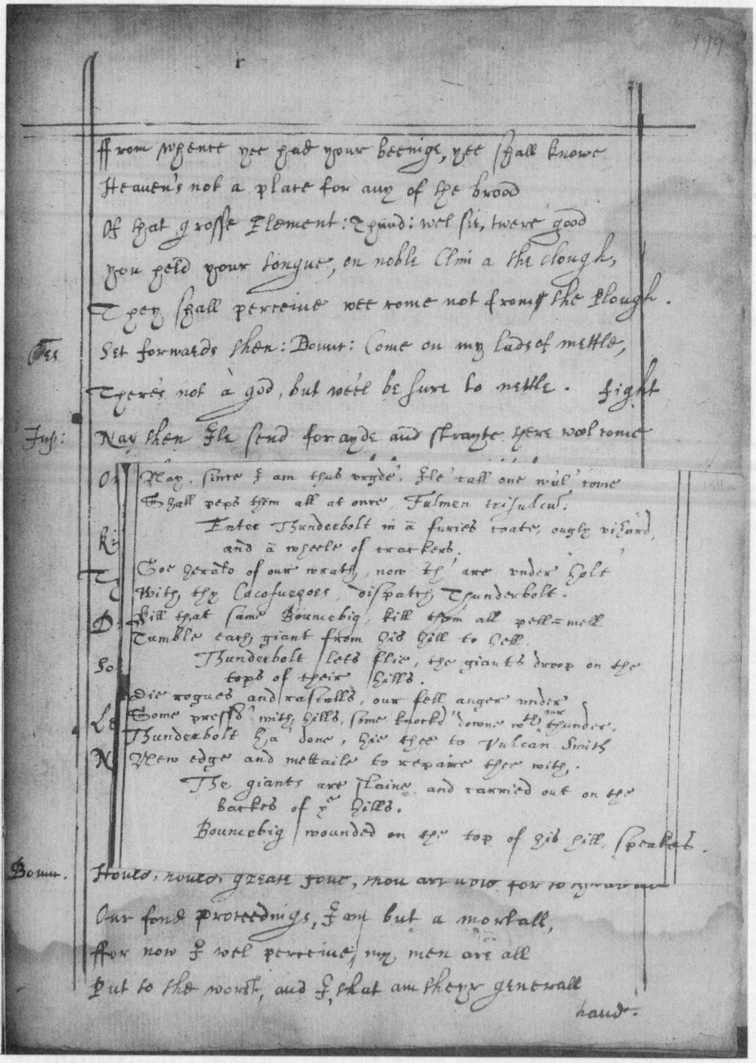

199

From whence yee had your beeinge, yee shall knowe
Heaven's not a place for any of the brood
Of that grosse Element: Thund: wel sir, twere good
You held your tongue, or noble Clim a the clough,
They shall perceive wee come not from the plough.
Ter: Set forwards then: Boun: Come on my lads of mettle,
There's not a god, but wee'l be sure to nettle. fight
Jup: Nay then Ile send for ayde and straight there wil come

Nay, since I am thus urgde, Ile call one wil come
Shall pepper them all at once, Fulmen trisulcum.
Enter Thunderbolt in a furied coate, ougly visard,
and a wreathe of crackers.
Goe herald of our wrath, now th' are under holt
With thy Cacofuegoes, dispatch, Thunderbolt.
Kill that same Bouncebig, kill them all pell=mell
Tumble each giant from his hill to hell.
Thunderbolt letts flie, the giants droop on the
tops of their hills.
Die rogues and rascalls, our fell anger under
Some pressd with hills, some knockd downe wth our thunder.
Thunderbolt tha' done, hie thee to Vulcan Smith
New edge and mettaile to repaire thee with.
The giants are slaine, and carried out on the
backs of ye hills.
Bouncebig wounded on the top of his hill, speakes.
Boun. Hould, hould, great Jove, thou art able for to hinder
Our fond proceedings, I am but a mortall,
For now I wel perceive, my men are all
Put to the worst, and I, that am theyr generall
hand.

PLATE 4: FOL. 199a, WITH CANCEL (TWO-THIRDS FULL SIZE): *GIGANTOMACHIA*, LINES 539 TO 562

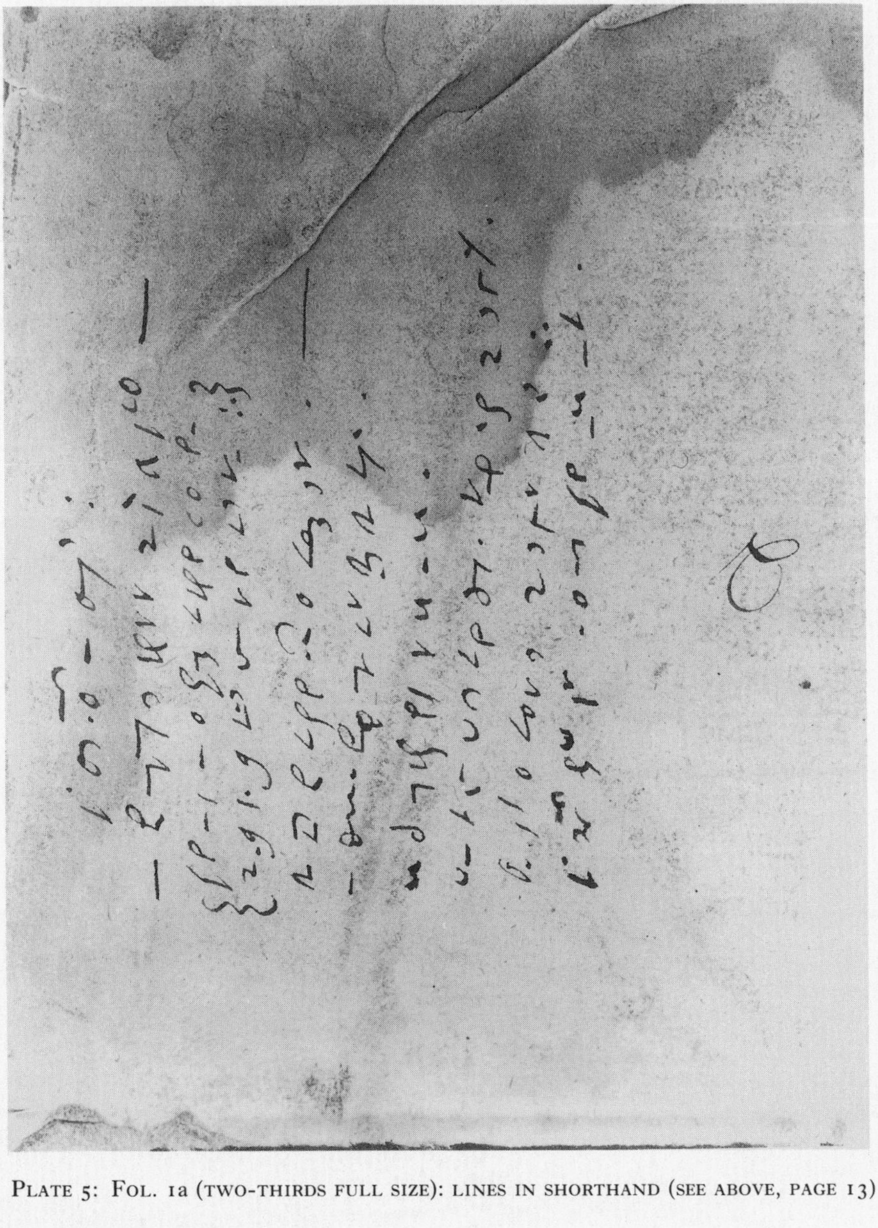

PLATE 5: FOL. 1a (TWO-THIRDS FULL SIZE): LINES IN SHORTHAND (SEE ABOVE, PAGE 13)

PLAYS FROM J.a.1: TEXTS AND HEADNOTES

Item 3 in J.a.1: Fols. 19a–23a. ***Boot and Spurre***

Boot and Spurre is the briefest of the J.a.1 plays, only 355 lines, with a cast of seven. The casting of 'Boote / a Trauelor' and 'Shooe / a Citizen' implies a possible contention for superiority, of the kind in the J.a.2 piece, *Ruff, Band, and Cuff*, but in fact the play recounts an essentially amicable meeting at the inn, 'ye signe of ye legge' where Slipper is Chamberlain. The prologue is Shoemaker, 'a shooemaker is the Show-maker'. The play's genre might be defined in the same way as that of *Ruff, Band, and Cuff*, a dialogue '*acted in a shew*'. The speeches are characterized by constant puns, scatological language (especially in the flyting exchanges between the servants Shoetie and Pump), and some satire on puritans. Reference is made to the travellers Sir John Mandeville, Drake, Cavendish, and most particularly to Coryat's *Crudities*, which is interestingly criticized for its excessive precision in noting every tree and gallows. The references to Coryat date the play as written after 1611 (when the *Crudities* was published). The *terminus ad quem* may be 1617, the date of Coryat's death.

Bentley notes that *Boot and Spurre* 'does not have the clear college and university allusions' of *A Christmas Messe*,[1] and it is possible, particularly in view of the diminutive size of Shoetie (a 'little goe byth' ground' (l. 145), 'my little Minum' (l. 298)), that the play was written for a school, or for pages in a great house. Hilton Kelliher points out, however, that *Boot and Spurre* and the J.a.2 plays *Ruff, Band, and Cuff* and *Gowne, Hood, and Capp* are similar to the tract, *Worke for Cutlers. Or, a Merry Dialogue betweene Sword, Rapier, and Dagger. Acted in a Shew in the famous Vniuersitie of Cambridge* (London, 1616). *Worke for Cutlers* does not have specific college allusions. Kelliher speculates that all four pieces belong to the period between 1613 and 1615, and he concludes that, although contemporary regulations governed academic dress at both universities, and the Inns of Court, none the less 'the weight of evidence tends to suggest that all these plays are connected with the University of Cambridge'.[2]

The booklet of *Boot and Spurre* measures 149 × 197 mm. It has neither ruled margins nor running heads. The speech prefixes begin approximately 20 mm indented, with the text indented 35 mm. The hand is primarily secretary, using italic for the epilogue and the occasional Latin word (*nil vltra*). Secretary is used for the speech prefixes. The hand is clear, although there is some uncertainty about the distinction between *m* and *M*, and raised letters vary in

[1] Bentley, v. 1295 (work cited on p. 3, n. 1).
[2] Private communication to the editors.

height (wth, M^{rs}). Apostrophes are placed carelessly, and speech prefixes frequently fall beneath the line. One special problem is the division of words such as 'somethinge' (l. 112) or 'whatsoeuer' (l. 124). This hand has two different forms of majuscule *I*, but does not distinguish them as *I* and *J*. At l. 110 another hand cancels 'drinke' and writes 'soake' above it. There are few stage directions, and no indications that the manuscript was used for acting. The nature of the corrections suggests that this is a scribal copy.

Boot & Spurre: [FOL. 19a]
&c.

Prologus } a Shoemaker
Epilogus }

Actors

Boote/ a Trauelor
Spurre/ Bootes man
Pumpe/ Bootes ffooteman
Shooe/ a Citizen,
Shootie Shooes page
Slipper a Chamberlaine

Prologus

Gentlemen you must vnderstand y^{t} a shooemaker is the Show=maker I haue beene these 2 dayes makinge a payer of bootes, Shooes, Pumpes, and Slippers, and they had not beene done yet but that I went through [stitch] stitch with them. These bootes my thinkes should fitte any man but those that are to bigge in y^{e} calfe these shooes will pinche noe man but those y^{t} are troubled with cornes These Pumpes will fitt any man but those y^{t} houlde dancinge to be suꝑstition These Slippers I am sure will please any man but those that are out at the heeles Howsoeuell Ile trie, for Ile goe— presentlye and pull them on: onely thus much let me speake in myne own defence

0.1–1 *Boot & Spurre: / &c.*] in Hand X (see above, p. 13) 4 *Actors*] lined between ll. 4 and 5 11 *Show-maker*] *S* resembles a *C*; *er* partially obscured by margin 13 *yet*] *y* heavily inked 15 *those*] *o* corrected from ?*e* 17 *houlde*] *h* obscured by ascending loop of *s* in *please* (l. 18) 19 *Howsoeuell*] *sic*

Spectators

Muse not we speake of such meane thinges as these
Our shallowe witts wade not aboue the knees
These lowlye thinges I hope will not distaste
Men of our trade goe not beyond [thes] there last

Enter Pumpe

Pumpe He y^{t} hath serued in the office of a footeman as longe as I haue done, though he be neuer so light at the first, ere ye yeare goe about will be a heauy man: sure I am borne to runne as long as I breath; but if the way had not beene so durtye I would haue showne my M^{r} a faire paire of heeles: for if euer he sende me againe to runne such away, Ile runne quite away. I haue come to day further then I hope I shall goe to morrowe. for my Mr Boote will be sure to haue his liquor ere he goes out of the Towne: My M^{r} hath sent me before to ꝓuide his Inne, but Ile meete wth him for it anone. I must to ye signe of ye legge there is my old ffrend slippar; I must needę to him: for Pumpe hath allwayes beene very inward w^{t}h slippar: The signe [FOL. 19b] of ye legg is either in Shooemaker rowe, or Hosier-lane I shall know it by this Its a legg in printe I see by good fortune I haue stumbled vpon the Legge at ye first in stepp. What so ho Chamberlaine

Slip: Anon Anon S^{r} (Intus)

Pumpe Its ye tricke of all these Chamb: neuer to doe any thinge at ye first word. what Chamber: come away and be hangd'

Enter Slipper

Sl: Anon Anon S^{r}/ what myne old companion Pumpe whome I dwelt so longe w^{t}h, let Slipper embrace thee

Pumpe Old Slipper I am glad to see thy heeles at libertie but Ile tell thee what, my Mr Boote is by this tyme at the next Towne

Sl: In faith ye shall be welcome, he is a very good ghuest and will

26 *[thes]*] *s* is a single downward vertical stroke 28 *Pumpe*] between ll. 27 and 28 31 *breath;*] perhaps *breath,* 33 *sende*] minims unclear; perhaps *seude* (i.e. *sued*) 40 *Hosier*] *si* resembles *st* 44 *Pumpe*] between ll. 44 and 45 45 *hangd'*] *sic*

come of roundlye, and therefore Slipper will be allwayes ready
for him
Pumpe But come lets see a chamber p^{r}sently
Sl: I p^{r}sently I knowe Mr Boote is a neate gentleman and therefore
he shall haue a handsome chamber p^{r}sentlye. Exit
Pumpe Make haste againe/ I neuer felt my selfe so weary in all my life
I perceaue nowe poore Pumpe y^{t} thou art euen past the best such an
other Iourny as this will weary thee quite out

Enter Slipper wth a Iugge of beere

Slip: I know thou art weary Pumpe come followe me into this roome:
heere Pumpe Ile giue thee a slashe (Drinkes)
Pumpe Doe thy worst, Pumpe [I] will neuer refuse good liquor
Me thinkes both our trades be like one an other: for both Chamberlain⟨es⟩
Sl: and ffootemen get theire liuinge by runninge vp and downe
Pumpe Nay soft a footeman hath ye better place for he may goe before his
M^{r}
Slip: Why are not Chamberlaines men of greate callinge? and euery one
y^{t} hath to deale wth them shall finde them men of good reckeninge
Pumpe ffaith Slip: there you came ouer me: But besides all this footemen
are very ꝓuident, for be theire M^{rs} neuer so wise they must rise
betimes if they will ouer reach them:
Slip: ffaith and Chamber: are as cunninge, for they haue all wayes a
tricke to deceaue theire M^{r}s: if he be aboue, they vndermine him,
if belowe, they be sure to come ouer him:
Pumpe Well Slip: Pumpe will not fall out wth you: weel' drowne our
quarrell in this glasse of drinke
Slip: What Mother Geeries [ro] rounde, one to an other:

Enter Boote and Spurre

Boote Come Spurre where art thou?
Spurre Harde at your heeles
Boote I doubt but for thee [Sl] Spurre I hardly should haue gott to my [FOL. 20a]
Iournyes end to night

65 *Sl:*] should prefix l. 64 speech 74 *vndermine*] perhaps *vnder mine* *him,*] perhaps *him;*
82 *[Sl]*] scribe carries *Sl:* prefix from the preceding lines into this sequence with Spurre

Sp: ffaith M^{r} we were faine to pricke on hard: Spurre did his best to sett you forward

B Well sayd Spurre, I see thou art all mettall

Sp: And yet as soone as you are at your Iournyes end you are ready to cast me of

B But yet thou hast the witt to sticke close vnto me

Sp: I doe it for your owne sake, for you can neuer ride with out me

B This is one thinge Spurre I knowe thee thrustie for many times thou hast fought in blood vp to the eares for my sake

Sp: Nay I would haue you knowe my seruice on foote to, for vnles I attend you, you can walke in no fashion

B But Spurre one thinge I like not in thee, thou spoilest all my horses, thou art a very soare rider:

Sp: Why I am as good as prouender to your horses, I will sticke close to theire ribbes

B Well Spurre let it be your parte to see my horses well curried while I goe call for my lodginge (exit Spurre) what Chamber

Pumpe Oh ther's my Mr Boote I knowe by his creekinge: Slipper now shewe you a Slippery tricke and slippe away ye Iugge, and I will pumpe for an answeare for my master.— (Exit Slipper) oh Mr Pumpe stoopes as lowe as your foote to welcome you to ye signe of ye legge

B: My thinkes Pumpe you are somewhat fine: I doubt somebody hath trimd you since you came in

Pumpe I must needes say your poore vassaile hath beene beselinge and I haue beene wth Slipper/ Slipper and I haue sett foote to foote and drunk⟨e⟩

B I these are your old trickes, as longe as there is any stronge beere to be had, Pumpe will [drinke] soake no water: but what haue you ꝓuided for my supper?

Pumpe. ffaith I haue cleane forgotten: I spoke for somethinge, but I know not what, call vp the chamberlaine, he knowes more of your minde

B Just such a tricke you serued me at fflushinge when you was drunke

87 *Sp:*] between ll. 87 and 88   91 *thrustie*] *sic*   97 *prouender*] ²*r* heavily inked   103 *master*] italic *m*; perhaps *Master*   108 *drunk⟨e⟩*] *u* may be an imperfectly formed *a*; *e* partially lost   110 *soake*] interlined, in a different hand and ink, above deletion

P M^{r} if you will beleeue me I was as sober then, as I am nowe

B Tis well you can be drunke and stande to it/ what Chamberlaine

Enter Slipper:

Slip: Anon Anon S^{r}: what is your Worships pleasure?

B What did my man bespeake for supper?

Sl: Three Iugges of beere and haulfe a dramme of Tobacco

B S^{t} S^{t}

Sl: O I crye you mercy, I mistooke the roome

B Doe you not knowe then whats ꝓuided for my supper?

Sl: Not very well S^{r} but whatsoeuer it be, it shall be ready for your worship: but you shall be [sure] sure of a messe of sallet oyle and a dishe of [FOL. 20b] freshe Butter:

B Very good

Sl: Nay it shall be very good

B In good time

Sl: And you shall haue it in good time

B Nay good Slipper stand not so much vpon your pantables, but answeare me soberly to one question, doe you not knowe one Mr Shooe?

Sl: Yes S^{r}, there are two of the name, one of them is a high Germain⟨e⟩ and the other a lowe Countryman:

B I meane him y^{t} dwells in ye Towne

Sl: I knowe him very well, he is a man of good fashion and a continuall howsekeeper

B And I heare good reporte of him, y^{t} Mr Shooe is a very vpright man:

Sl: Yet for all y^{t} heele goe awry sometimes, and this is his fault he loues to be greacd' often. but as I take it heere comes the gentleman:

Enter Shooe and Shootie

Shooe Stringe come tye my Shooe

Pumpe What little goe byth' ground is that

124 *whatsoeuer*] perhaps *what so euer* 145 *Pumpe*] vertical stroke in *P* heavily inked; scribe began with *B* *byth'*] apostrophe present but inked over by *t* and *h* of *byth* and tail of *y* in *my* (l. 144)

Slip: A hangeby of Mr Shooese, they call him Shooetie: a fellowe I am not behouldinge to: but farewell a while Pumpe

Pumpe Slipper farewell, Pumpe will haue a single boute anone w^{t}h you

Shoe Sirrah how does my cosen Stockinge

Shooetie ffaith S^{r} he was lately a dyinge, but now hees on ye mending hand

Shoe Why what was ye matter?

Shooetie They say he tooke a greife be cause he stretched himself so farre for Mr Legg that he broke

Shoe Well if y^{t} be all weele haue the matter taken vp: but for my neighbour ffoote what sayes he?

Shooetie Marrie ffoote says he skornes to be kept in by you and vowes as he is a parte of ye body politicke and a member of ye com̃on wealth y^{t} he will stand in opposition against you, and doe his best to treade you vnder

Shoe Alas poore wretch I tooke him in when he might haue stood and coold—— his toes, but tis no matter let him goe whether he will he neuer comes w^{t}hin my latch [against you and doe his best to] againe [treade you vnder:] : but is not yonder Mr Boote?

Boote Mr Shooe

Shooe Your poore frend Shooe will allwaies be ready to doe Boote anie seruice one y^{t} will euen spend life and soule to doe you any good But my thinkes Mr Boote you looke somewhat old, for you are very ffull of wrinkles [FOL. 21a]

Boote Age and trauell will alter men

Shoe I hope Boote hath. beene no greate Trauellour

B As greate as any y^{t} goes vpon earth for trauell Coriat could neuer come neere me

Shoe I thinke so, for Coriat went all wayes on foote, and you trauell all wayes on horsebacke: but what thinke you of Coriats booke of his trauells?

Boote Surely I thinke he tooke greate paines and trauelled hard for it, but his Crudities—I could neuer digest them, he cannot passe by so much as a

146 *Slip:*] between ll. 146 and 147 161 *coold*] ^{2}*o* altered from *a matter*] perhaps *matter-* 162 *againe*] interlined above deletion 165 *allwaies*] perhaps *all waies* 170 *hath.*] period, a stray mark? 173 *all wayes . . . all wayes*] perhaps *allwayes . . . allwayes*

wallnutt tree but he needs must haue a flinge at it, he can as well be hangd as passe by a gibbet and euery paire of gallowes he renownes as if they were as famous as Hercules pillars with *nil vltra*: but because he is my fellowe trauellor I let him passe

Shooe But what say you to Sr Iohn Mandeuill they say he was at the worlds end, and then I am sure none of you can goe beyond him.

B I thinke for Linge, for he will lye to a haire, he goes and looses the haire of his head in some hot skirmish and then says it was burnt of wth goinge to neere ye Sunne

Shooe ffayth yt was but a bauld excuse and yet it was well enough put of. but lets leaue these land trauelors for a companie of drie fellowes, now for water men, lets haue an oare in theire bote, what say you to Drake and Candish

B ffayth for Candishe he made a good voyage at the first but at the second he had so farre engaged himselfe yt he was cleane ouer eares here was a waire and soe in ye end he proued a castaway

Shooe Oh tis an excellent thinge in a Marriner to keepe himselfe wthin compass⟨e⟩ but how sped Drake?

B Drake at ye first was but a wild fellowe but before he had done he soe trounced ye Spaniards yt he made them Quake Quake

Shooe And therefore me thinkes seamen of all men stand best to theire tackling but whats ye reason these puritans cannot abide them

B I thinke it is because they hate the Sea of Roome

Shooe Or it may be because they vse ye Cardinall windes

B Or ꝑhaps because they detest the inuention of Cardes

Sh: Or else it is because they haue allwaies a Pilate in theire companie

B And I am partly of theire minde, I haue gon as farre as any man and yet by my good will Boote should neuer take water: but by this tyme Mr Shooe I thinke your heeles doe ake wth standinge & therefore sitt downe heere and Ile vnfould vnto you the whole discourse of my Trauells in priu⟨ate⟩

Shooetie Now whilst my Mr and Mr Boote are setled Ile enter company wth this

177 *needs*] *s* altered from *e* 184 *skirmish*] *k* altered from *t* 193 *compass⟨e⟩*] *e* lost in cropping 197 *tackling*] tail of *g* lost in cropping 202 *allwaies*] perhaps *all waies* 207 *Shooetie*] between ll. 206 and 207

younge gentleman S^{r} your ffrend Shootie had no sooner got loose but he needes must salute you in ye way of kindnes

Pumpe S^{r} in reciprocall congratulation I kisse ye very shaddowe of your shooesstring

Shootie Pray you S^{r} let me craue your name [FOL. 21b]

Pumpe Doe then

Shooetie Then w^{t}h your good leaue and likinge whats your name

Pumpe My name S^{r} is Water Pumpe

Shootie Very well water: I thought all this while your name was water you looke so thinne ont, surely he y^{t} made your Trouses was a very good Taylor, for he went roundly to worke

Pumpe Why sirra Iack sace I hope I may weare these Nockes in spite of your nose

Shooetie Surely you are a very fortunate man, you were lapt in your mothers smocke for she hath giuen you one of her old blankets

Pumpe Shooetie you harpe to much vpon one stringe but though your tongue be neuer so busie about my nockes I would not haue you thinke you can put downe my breeches:

Shootie Shite on your breeches I care not a fart for them

Pumpe Well I perceaue Shooetie if I should stay neuer so longe youle neuer leaue wagginge therefore Ile leaue you to my fellowe Spurre

Enter Spurre w^{t}h a Cloake-bagge

Spurre Good Ostler if thou euer meanst to currie fauor w^{t}h me looke well to my horses, and let them be well littered, or else I shall not care a strawe for thee, doe not play ye fresh seruingman w^{t}h them, & take away theire meate before they haue done but lay downe theire ꝓuender before them, and let them trye it out by the teeth

Shooetie Mr Boote is a straunge man he hath a paier of seruingmen y^{t} looke as though they were not fellowes, one is in his nockes and ye other in his cloake-baggs: pray you S^{r} before you goe, giue me leaue to Spurre you one question

Spu: S^{r} if you will giue me leaue to kicke you an answeare

Shootie Pray you S^{r} what countryman are you

210 *shooesstring*] *sic* 220 ^{2}*you*] *u* written over two letters, perhaps *ur* 225 *Shootie*] between ll. 224 and 225 230 *my*] *m* blotted ? 231 *seruingman*] ^{2}*n* blotted 235 *one*] *e* blotted

Spu: A Northhamptonshire man, I was borne at Rowell

Shootie I had though you had beene a Germaine you are somewhat a kinne to Kickerman

Spu: I perceaue shooetie you are a notable Slipstringe but Sirrah is thy M^{r} heere?

Shooetie What if he be, Spurre hath nothinge to doe wth him

Spur: Oh yes he is come in very good tyme, for all my horses want shooses

Shooetie Doe not disparage my M^{r}, I will not chaunge wth you w^{t}hout Boote

Spu: Nor I with you for all the horses in your shopp

Shooetie Goe to hould your peace for if Boote be to lustie weele haue him strapt

Spurre Nay boy then hould you your pratinge if Shooe flie out Ile clout him my selfe, what is thy M^{r}, he is but a Citizen, he is but a prentice at ye best

Shooetie And thy Mr Boote is a trauellor, and therefore he is but a Iournie-man

Sp: Shooetie hould your peace or else I shall hange you vp for the signe of the rose [Fol. 22a]

Shooetie I can tell you y^{t} my rose is ye best flower in my M^{rs} garden, but I wonder Spurre why you are so sharpe, for all your greate sword if I once buckle with you Ile lether your hide

Sp: I hope my sword doth not trouble you, if you be to busie Shooetie, take heede for this is ye greate sword y^{t} cutt the Gordian knott

Shooetie I could neuer tell ye reason why euery such clownish Seruingman is tide to such a sword, vnles it be y^{t} when they farte they may quickly touch could Iron:

Sp: Sirrah Ile tell you wth ye pointe Ile pricke a hole in your coate and with the hilt pum̃ell you about ye pate

Shooetie But stay Spurre heere comes your M^{r} to take you offe

Boote Mr Shoe you and I come both from one howse for as Antiquaries saye we are lineally descended from Thonge Castle:

241 *though*] *sic* *somewhat*] perhaps *some what* 267 *the*] *th* heavily inked

Shooe Stay there Mr Boote, I thinke you be wide for I haue heard some hould y^{t} our aunchestors were borne at Buluerhide, but its no greate matter, its certaine we came both of one stocke

Boote But I forgot to tell you one thinge, as I passed alonge this towne side (he looks on his mapp) I had much adoe to scape this tree, and so com̃inge along ffraunce to Deepe, I fell vp to ye topp in a deepe hole, and was so bemired, y^{t} I was fayne to get an ostler to make me cleane:

Shooe By Lady S^{r} it was a fowle mischaunce

Boote Nay I can tell you Boote hath had many of these Iournyes, for he hath gon through thicke and thinne

Shooe Well now I see plainely before my face y^{t} you are a greate trauellor you excell Vlysses himselfe for all he was so longe abroade y^{t} when he cam⟨e⟩ home his wife knew him no more then his dogg and Æneas though he w⟨as⟩ a true Troian yet came farre [yet] behinde you

Boote As for Vlysses Circe made a hogrubber of him and for Æneas any asse may doe as much as he did

Good S^{r} let me but intreate one thinge of you that you would bestowe

Shooe this mapp on me

Boote ffaith Mr Shooe Ile be lought to sticke wth you for anythinge, but you must pardon me I would not parte wth this mapp for a world: but I thinke I haue held you to longe, and therefore to winde vp all in a word after this raw discourse to stop your mouth I envite you to parte of a badd supper:

Shooe No faith S^{r} I am to goe an other way

Boote Nay Mr Shooe I will not be put of so

Shooe Then good Mr Boote leade you ye way Ile follow at an inche

Exeunt Boote & Shooe

Shootie I hope S^{r} you are somewhat cooler you are not a hot spurre still

Spurre No faith S^{r} my little Minum it was but a flashe but why went my fellowe Pumpe away

275 *adoe*] perhaps *a doe* 288 *Shooe*] should prefix l. 287 298 *Spurre*] between ll. 298 and 299 *it was*] closely spaced with a small *i*; perhaps *,twas*

Shooetie It seemes he had beene drinkinge somewhat stiffe and so your fellow [FOL. 22b] Pumpe could not hould his water

Enter Slipper w^{t}h a Iugg of beere & a glasse in his hand

Spurre How now slipper neuer w^{t}hout a Iugg of beere in your hand

Slip: I heere you may see as in a glasse ye life y^{t} we chamberlains doe leade

Shooetie Come tis in a good hande pray thee begine to somebody

Slip: You may be sure Sirrah Ile not begin to you for I was neuer bound to you Shooetie

Shooetie And y^{ts} it Slipper hath made you such a loose fellowe

Slip: [Y⟨.⟩] goe to Shooetie you are a notable wagtaile: heers to thee Spurre thou art all steele to ye backe

Spurre Come on Ile neuer refuse Chamber: but I am haulfe out of loue with these Ostlers, one of them laste night gaue me such a wipe

Slipper drinkes

Shooetie What a whole one

Sp: What a whole one, what else slipper was allwayes a drye fellowe

Shooetie And I haue allwayes markt this in him y^{t} neuer while slipper is sober he is in ye plaine fashion, but when he is tipsie he is allwayes in his turneouers:

Slip: Well sayd Shooetie if you vse these trickes longe you may chaunce to come to be knitt vp: but I pray you S^{r} take your liquor

Sp: Heere Shooetie (drinkes) in truth I wonder at these tapsters how they can beare such a deale of drinke such an other glasse as this would make Spurres head runne round:

Enter Pumpe

Pumpe I though were I should finde you, you must stand beselinge heere whils⟨t⟩ poore Pumpe must dance attendance come away quicklie or else my M^{r} sweares heele haue you all sett in ye Shooemakers stockes:

Shooetie ffaith for my Mr Shoe I care not, I haue him in a stringe

302 *&*] very weakly inked; perhaps *(* 312 *these*] ^{1}*e* badly formed, perhaps *o* 315, 316, 317 *allwayes . . . allwayes . . . allwayes*] perhaps *all wayes . . . all wayes . . . all wayes* 325 *though were*] *sic* 328 *in*] *i* altered from *o*

Spurre ffor my Mr Boote I haue serud him of and on a good while wherefore now he may forbeare me a little:

Shooetie ffayth Pumpe my thinkes thou stanst as though thou hadst neither life nor soule, wilt thou drinke wth vs Pumpe

Pumpe Noe by noe meanes, I had as leaue you should powre it in my shooses

Sp: I hope Pumpe you may be drawne on by a little and little

Pumpe Nay if you be there abouts I must take my heeles/ Exit

Sp: Though I care not for my M^r, yet I loue my supper therefore lets goe in and after supper weele haue a catch in ye chimnye corner

Shooetie Content/ Exit Shooetie

Slip: Pray you S^r while you sing a songe let me beare ye burden (takes ye Cloakebagg) Thanke you hartilie tis for my owne ease I hope you haue neuer a cloake for your knauerie for if you haue you haue giuen me the bagg handsomely

Exeunt Slipper and Spurre

Epilogus a Shooemaker wth an end in ye one hand and an all in ye other

Shooemaker Gentlemen y^e ꝓlogue is now returned at ye last[t] wth a Shooemakers end, this is all

If bungler like my worke be brought to end [FOL. 23a]
Ime but a Cobler who you know may mend
Next tyme I hope my worke shall be more meet
Now I haue learnd ye lenght of all your feet
But if in workemanship I doe excell
I know your bountie, you will pay me well
Or else Ile trust and bookt at your commaunds
So that you will subscribe it with your hands

finis

333 *powre*] heavily inked *w*, perhaps covering an earlier *u* 335 *heeles*] *l* heavily inked, perhaps written over *t*

Item 10 in J.a.1: Fols. 105a–115b ***A Christmas Messe***

A Christmas Messe, dated 1619 in the manuscript, is the most specifically collegiate of the dramas included in J.a.1. As is made clear above (p. 3), the play is not a salting, although it contains a reference to one. Bread says to Salt (ll. 76–8):

> For now thou art not only for the table,
> But for another stinking, countless rabble,
> I mean the Freshmen . . .

If at this time the word 'sconce', both noun and verb, was used exclusively in Oxford, as David L. Russell argues, then *A Christmas Messe* must come from an Oxford college. Bread tells Salt that the Butler was 'sconct for spotts' (l. 105), and Vinegar claims to have 'sau'd the Cooke a sconce' (l. 412).[1] The occasion is the Christmas season, although after Christmas day itself: Cushion talks of 'th'inraged fire, that's full as hot / As was at Christmas our plumporridge pott' (ll. 114–15). It is the evening on which guests are entertained: 'to night . . . strangers must come / To make their Chrismas feast, as is their custome' (ll. 126–7). The Cook, as epilogue, again recalls the occasion: 'y' 'aue rime, & though not good yet 'tis in season' (l. 674).

The play was probably performed in the college hall (Belly comes 'into this goodly hall / To find good cheare', ll. 7–8), and the staging was adapted to an after-dinner performance. At the end of the first act, Salt, Bread, Cushion, Tablecloth, and Trencher all 'take their places vpon the Table' (l. 132). Besides the table the only props needed are those used in the final battle between the forces of King Beef and King Brawn: 'Spitts, & dripping / pannes, and such like weapons' (ll. 583–4). As in *Heteroclitanomalonomia*, there is only one female part, in this case Queen Mincepie.

The college atmosphere is indicated by the emphasis on food, the scatological jokes, the references to the college cook, laundress, scholars and fellows, and the daily menu (fish, beef, and an occasional rotten egg). The freshmen, the stinking countless rabble, are teased for eating bread. The town and gown hostility surfaces when King Beef, asked what he will do if he cannot find his father's horn to use as a weapon, reassures Vinegar with the question (ll. 354–56):

[1] Russell's evidence is discussed above, pp. 3–4. Further doubts about the Oxford identification come from Hilton Kelliher, who compares the joke about cleansing the freshman at the expense of 'Sr Harry Bath' (l. 79) to the Cambridge coinage '*Harry Soph*', 'said to have come from the reign of Henry VIII and denoting a scholar who has completed four years' study towards his B.A. but has declined to take his degree' (private communication to the editors).

Are there not horned beastes enough on grownd?
Or if the fields bee scant, & can not fitt yee,
yee may finde hornes to many in the citty.

A Christmas Messe also contains many classical allusions, and, as Belly assures us (l. 4), it is no simple 'mockshow'. An entire five-act structure is squeezed into its 679 lines. Further, when Belly tells Cook that Beef and Brawn have gone off to battle, he adopts a mock-heroic manner (ll. 538–40):

Thou great Apollo, who's sweet harpe is strunge soe well,
Guide mee now speaking, that cann'st rule the tounge soe well
A bloudy quarrell must by mee describ'd bee.

The debate plot is based on an argument between King Beef and King Brawn for precedence at the Christmas feast. Beef 's lieutenants are Pepper and the cowardly Vinegar, Brawn's are his son Sauce and Mustard. There are flyting scenes as well between Tablecloth and Trencher (I.i), and Bread and Salt (I.iii). The resolution is provided by Cook, who uses his chopping knife rather as Lily uses the rules of grammar in *Heteroclitanomalonomia*—and with a final threat like that of Thunderbolt in *Gigantomachia*.

The manuscript of *A Christmas Messe* measures 150 × 193 mm. The title-page is ruled at the top and on the left side (see Plate 2). The left margin, which continues throughout the manuscript and contains the speech prefixes, varies from 15 to 20 mm. There are no running heads and no catchwords. The hand is primarily italic throughout, with variation for headings. The manuscript has suffered severely from show-through, which almost obliterates some words in the later leaves. The scribe uses both *v* and *j*. One characteristic spelling is 'Chrismas', which freely alternates with 'Christmas'. Question marks are reversed, and the ampersand is preferred to 'and'. There are no corrections in other hands.

A Christmas Messe: Actus 1ˢ: Sce: 1ª: 1619 [FOL. 105a]

Enter Belly:

Bel: S't, what's the matter? why doe yee flock soe?
yee thinke belike I'm prologue to some mockshow.
In this yee neyther wise men are, nor witches,
If yee thinke soe, beleeu't yee wronge your breeches.
For I am come into this goodly hall

1 *1619*] added in Hand X (see above, p. 13)

To find good cheare, & soe I hope I shall.
For wot yee who I am? Belly's my name
A man I'm sure this Christmas in good fame.
Wer't not for mee, what would your victualls doe?
Euen lye & stink, & mould, I & worse to:
How many Butchers, Bakers, Grosers, all
To Belly to deuoure apace doe call.
If I but once grow queasy, all their ware
Growes straight as cheape as 'tis at Bartholmew fare.
I'm only in request. for who not wishes
A Belly correspondent to his dishes?
And now I hope to stuff my gorrell full,
This Christmas. But this Cooke, this greasy gull
Soe vexeth my poore heart with expectation,
That I could eate him vp without compassion.
well Ile goe call him. why doe yee looke after mee (hee lookes
Beleeu't I did not come for you to laugh at mee. back.
Exit:

Scena: 2^{a}:

Enter Trencher and Tablecloth.

Tren: Come Tablecloth, heer's such adoe I wisse
'Twere time iffaith you had been layde ere this.
Tab: Faire S^{r}, you are as briske as 'twere a wencher,
Ere dinner's done, you'l bee a greasy Trencher.
Tre: And thou foule Tablecloth, our fortune's one, [Fol. 105b]
wee shall bee in like pickle both anon.
Tab: And yet how square thou sitt'st, fie 'tis not good,
when all men know th'art but a man of wood.
Tr: Nay Tablecloth, if once you goe to flouts,
who knowes not you to bee a man of clouts?
Tab: yet my birth's better, I at first was faire,
Thou but a rude chip, till they made thee square.
Tr: Alas dost bragg of that, th'art yet to seeke,

24 *back.*] between ll. 23 and 24

Poore foole th'are faine to wash thee euery weeke
Tab: Nay Trencher, yet for all thou hast not scapt,
'Tis better to bee washt, then to bee scrapt.
Tr: That proues my strenght, but thou poore ragg of draping
Thou art soe thin, thou canst not hold out scraping.
Tab: Thin am I? hold thy peace, yet th'art noe winner,
Ere longe thy scraping soe will make thee thinner.
Tr: Thinner then thee! fie cease these idle braggs,
Thy washing will weare thee to rags, & jags.
Tab: yet when I'm soe, I serue for some good turne,
when thou art good for nothing but to burne.
Tr: Then thou wipst dishes, & tailes to I doubt mee,
whilst I giue heat, & comfort all about mee.
Tab: Nay but thou burnst, & soe to dust art brought
And when thou art soe, th'art just good for naught.
Tr: Nay when I'm ashes Tablecloth I'm better
For then I wash thee, still thou art my debter.
Tab: wash mee? why then thou seru'st mee, this is braue
Here take thy wages, speake what wouldst thou haue?
——But here come Bread & Salt, let's stand alofe
Anon wee'l vrge this matter to the profe.
Bee silent, whil'st wee heare thẽ make their braggs,
I thinke wee shall haue pretty sport effaggs.

Scena: 3ª: [Fol. 106a]

Enter Bread, and Salt:

Salt: Bread, prythy learne thou to come after Salt.
Bre: yes marry, soe I will when I am halt,
And cannot get before. Sa: why Bread thou know'st
That thou wast but a lumpe of past at most,
Vntill I season'd thee, & brought thee in
To great mens houses, & the Butlers Bin.
Br: Masse well said Salt, th'art very witty growne,
And yet noe maruaile to if all were knowne.

48 *thee!*] perhaps *thee?* 55 *naught*] *a* altered from *o* 58 *wash*] perhaps *Wash*

Since th'art soe cõmon made, thou can'st not chuse
But learne it in the company thou dost vse.
For now thou are not only for the table,
But for another stinking, countless rabble,
I mean the Freshmen, which ofte⟨n⟩ tryed hath
Beene to the cost of good Sr Harry Bath.
And beeing among such witty prating fellowes
(Vnles thou wer't as senceles as a Bellowes)
Thou canst not chuse but grow a great deale riper
In quick conceits, that passe Tarletons the piper.
Salt: Alas poore Bread thou art ten times more common
To euery Freshman, I nere saw thee frõ one.
Marke but a sneaking Freshman, when hee comes
Biting with hungry teeth his durty thumbs
Heere for his Beauer, with what courage straight
Hee cryes a loafe, a loafe: and if hee waite
Longer then pleaseth him, hee will begin
To scould and braue, & snatch thee from the Bin.
Noe place can hold thee frõ thẽ, th'art not able
To saue thy crũms left on the Bousers table.
If there bee any scraps of thee appearing, [FOL. 106b]
Thoughthey bee durty, & not worth the hearing,
Downe thier wide mawes 'tis tumbled; they soe loue thee
Though Shoogrease butturd thee they would approue thee.
Bre: Fie Salt thou stomack broyler, blame mee not,
If with thy wordes I grow outragious hot.
Is't a disgrace for mee thou fretting Elfe
To bee abusd, that cannot right my selfe?
But thou o plagy Curre, when noe man moues thee,
Willt worke thy mischeife, woe to him that proues thee.
The Butler that's thy Landlord, owes thee hate,
Oft was hee sconct for spotts vpon the plate.

78 *the*] *t* altered from *a* 95 *Thoughthey*] *sic*

Salt: well bread, I scorne to take these words th'hast vtturd,
Thou talk'st as nimbly now, as if th'wert butter'd
Bre: Leaue pratling quicly, some body doth rush in
Salt: Tush feare not man: oh 'tis our fellow Cushion.

Scena: 4^{a}:
Enter Cushion:

Cush: your fellow quothe? how can you bee fellowes,
That puff out anger like our Colledge bellowes?
Against th'inraged fire, that's full as hot
As was at Christmas our plumporridge pott.
If thou wert well serud Salt, thou wouldst bee beaten:
And thou Bread hast deserued to bee eaten.
And Trencher cause you kept noe better square,
The scrapings of the meate shall bee thy Share.
And Tablecloath, allthough thou feare not mee,
The Landresse fayth shall haue a bout with thee.
I thinke y'are all like bells, or else like bacon,
yee neauer will bee good till yee are taken,
And hang'd vp, then perhaps you'l cry Peccaui, [FOL. 107a]
Effaith Ile teach yee better to behaue yee.
Doe yee not know to night that strangers must come
To make their Chrismas feast, as is their custome?
Leaue, leaue, and doe not stand thus thrũming capps:
The messes will bee now sent in perhaps.
Then each man take his place about the table,
Tide with a bond of loue, as with a Cable.

They all take their places vpon the Table:
then Enter King Beefe, wth Sr Vinigar, & Sr Pepꝑ:

Actus: 2^{s}: Sce: 1^{a}:

K: Bee: Where are my Knights? where art thou Pepper to,
Where good Sr Vinigar? Ambo: All about you.
K: Bee: Thus it befitts King Beefes great majesty

106 *vtturd,*] *d,* added later with a sharper pen 114 *fire*] *re* altered 115 *plumporridge*] perhaps *plum porridge* 128 *thrũming*] *m* missing one minim

To walke the hall in state. Loe heere come I
Newly frõ fiery Phlegeton: ah yet
I feele the anguish of the iron spitt,
That late trans[x]fixt my body, still I thinke
The greasy Cooke bathed in sweat & swinke,
Larding my broyled corpes, & others by
Pinching my sides with double cruelty.
But yet those hands that late tormented mee
Those greasy hands at last haue set mee free,
And gi'n mee leaue to see this goodly many.
Can yee now find [ye] in your hearts good people, can yee
After my tortures yet more to abuse mee?
And mocke my to hard fortune? fie yee vse mee
Not like a King, but I will proue I am one,
Allthough the greasy Bores most mighty gãmon,
Brawne that imperious slaue doth pick a quarrel
Against our highnesse; when the old foule Barrell [FOL. 107b]
Is scarcely yet deliuered of the burthen,
And keene-tooth'd Mustard hanges vpon the lurden,
But now hee offer'd mee abuse i'th Kitchin.
Vin: And mee to. oh I felt my fingers itching
At the proud slaue, who though not halfe soe able
As is your selfe to furnish out a table,
yet would hee needs bee seru'd in first, but for it I
Will quicly crack his arm'd superiority.
Pep: Art thou not greater farre then Brawne or Souce is?
Sure not a greater King in all this house is.
Or that hath such a Queene to take his part⟨ ⟩
Queene Mincepy the abstract of all Cookeryes art. {Sce: 2ª: Enter
Soe well belou'd that many her doe vphord {Que: Mincepy
Frõ Christmas vnto Easter in their Cupbord
K: Bee: And heere shee is, how gloriously shee comes?

148 *Can*] *Ca* altered from *An* 166 *Sce: 2ª: Enter*] between ll. 165 and 166
167 *Que: Mincepy*] between ll. 166 and 167

Shee breaths out nothing but sweet spice & plũms.
See but how smooth & round shee's in the wast.
Her sides begirt with walls of solid past.
Come my [owne] faire Queene, mine owne deare flesh & matter,
Would not thy bewty grace a pewter platter?
And more beseeme our stately Christmas bord
Then clownish Brawne, or Souce that hoggish Lord?
wee'l scorne to yeald to their vpstart authority,
Or loose one tittle of superiority.
This rude rebellion farre more stomack I
Then did the Gods the Gygantomachy.

Q: Min: My royall Husband, now I see th'art valiant
And from thy Fathers prowes nothing aliant.
Let not their coward threatnings vs apale
What wee haue might and will to thumpe thẽ all.
Wee'l cleanse our hearts frõ sorrow if wee can,
As doth the dishclout the foule drippin pan.

Vin: Brauely resolu'd (deare Lady) for my part [Fol. 108a]
I doe not care (Saue reuerence) a ——
But will behaue my selfe as valiantly,
As ere did Warwicks thrice renowned Gye.
Hee kild the Boare, & I will kill King Brawne,
Which to performe I mine owne life will pawne.
when I haue killd him, then Ile teare his jawes
The slaue shall neauer scape our furious pawes.
And wth my speare Ile giue him such sore hunches,
That when heė's dead, Ile make him feele my punches.
I come with thunder Brawne, Dub; Dub, a Dub,
And if that faile, Ile kill thee with a Club:
And if perhaps that should not make thee dye,
Beeware King Brawne Ile kill thee, with a fly.

Pep: Sr Pepper vowes as much in's M^{rs} seruice,

170 *sweet*] ^{2}*e* altered from *a* 173 *my*] *y* altered from *in* *faire*] interlined above deletion with caret 180 interlined between ll. 179 and 181

Whilst in this microcosme one valiant nerue is.
wee scorne to take the least abuse of such a one
As that proude king, I'le giue him such a touch anon,
Shall make him know King Beefe hath seruants strong enough
To hold out tack with him, I and that longe enough.
Wee are full fead with solid Beefe & Mutton,
Tut, tut, wee care not for King Brawne a button.

K: Bee: O most Herculean spirit, Hector of Troy
we[e]re hee cõpar'd to thee, was but a boy.
And stout Achilles arm'd with rage & steele,
Whose valour (as they say) lay in his heele.
Now wee'l to parly faith, Brawne Ile talke w^{t}h thee,
Come sweete Queene Mincepy, will you walke w^{t}h mee.

Q: Min: With all my heart, and when the villaine rayles,
I'le plucke his venemous tounge out w^{t}h my nayles.

Exeunt:

Scena: 3^{a}: [FOL. 108b]

Enter King Brawne, and the Lord Souce
his sonne, Mustard his attendant:

K: Bra: Thus from the Souce=tub where in woefull plight
I haue laine bathing, come I to the light.
My deare Sonne Souce, & thou my faithfull Mustard,
How glad I see you: oh I haue beene thrust hard
In that darke dungeon, bound with hempen cord,
Rowld like a footeball, but am now restord.
I haue beene much abusd by great King Beefe
O hee's a notable tough & sturdy theefe.
But wee will not bee trode thus vnder foote
whil'st Mustard, my Sonne Souce, & I stand to't.
Thou art a King, then beare a Kingly spiritt,
And set thy selfe as high, as thou dost meritt.
Thou now art in the fauour of the people,

205 *enough*] *ugh* written above *eno* 207 *fead*] *sic* 224 *thrust hard*] perhaps *thrusthard*
227 *great*] altered from *grett*

As high I thinke in conscience as Paules steeple.
Then doe not let that proud King put thee downe
Feare not those paltry hornes that guard his crowne.
Ile keepe the custome still inuiolable,
Brawne shall bee brought first to the Christmas table.

Must: I marry shall hee. can it not suffice him
That all the yeare men doe soe highly prize him
But nowe to when wee only are in season,
Hee must bee the cheife man. Tut there's noe reason.

Sou: Giue him but once an inch, hee'l take an elle;
And thus 'twill bee, till wee his bowldnes quell.
Hee and his fellowes all shall know prince Souce,
when fit occasion serues, will proue noe Mouce.
Let vs prepare our armies mighty Brawne,
I' shall proue the bloudiest fight that ere was sawne.

K: Bra: I thinke thou art a witch thou read'st soe right,
I promise you 't will proue a bloudy fight.
Wee'l make the welkin roare, & Phœbus hee [FOL. 109a]
Shall come downe tumbling from his Appletree ⌜Axell tree⌝
The terrour of our weapons furious clangor
Shall moue great Iupiter himselfe to anger.
Thou Souce shallt bee our great Leiftenant generall,
I thinke more ready for the battell then are all.
A younge Porkes spearerib Souce shall bee thy speare,
Our selfe our owne shield to the field will beare.
Thus arm'd wee doe not feare (& 'tis noe wonder)
Should Ioue descend to fight with vs in thunder.
I hope thou'lt stirre thy stumps as liuely Mustard,
As the braue combatants at longe spoone & Custard.

Must: Doubt not my valour noble King, our foes
Ere they see mee, shall feele mee in their nose.
And as the Mouce doth kill the Elephant; Soe

243 Sou:] above line 249 *witch*] *t* altered from *c* 252 ⌜*Axell tree*⌝] written below *Appletree* 253 *clangor*] original *r* obliterated; *r* interlined above *g* 254 *Shall*] *S* altered

I meane to kill King Beefe before I goe.
I feare not Vinigar, nor spruce Sr Pepper
Nor yet Queene Mincepy, but if I once gett her,
I'le make her Queeneshipp know I am able to make
Her tumble downe into my Chrismas stomacke.

Sou: But first I thinke it fit, that wee goe finde
King Beefe out, and to know the rebells minde:
Perhapps hauing better thought on't, hee will yealde,
Before wee bringe our forces to the fielde.

K: Bra: Thou counsell'st well, and like our princely sonne
Wee'l make some hast to haue this battell done.
Least if the guests should finde vs in this fury
In their vnmercifull jawes they should vs bury.
Lett's goe to parly then, Mustard bee neere mee,

Must: vnlesse you wipe mee of, you neede not feare mee.

Scena: 4^{a}: Exeunt omnes.

Bread, Salt, Trencher, & the rest
stepp forth, and speake:

Salt: Now Bread you need not striue who shall bee greater, [FOL. 109b]
For if this discord holde, you'l want an eater.

Brea: And thou to. Table: But mee thinkes I'm ouerjoyd,
By this meanes I am kept still unanoyd.
They will not bee made freinds in hast I hope,
Soe shall my Landresse saue both paynes & Sope.

Tren: I am of thy mind Tablecloth: Cush: I doubt
Cushion will not soe soone bee worne out.
Thus when wee feast least, then doe wee fare best,
And in our propper Mansions take our rest.
But let vs watch a while, least vnawares
The Cooke should reconcile these furious jarres.
Lett's take our places: heere comes boundles Belly:
Hee smells a feast. Brea: S't by and by hee'l tell you.

They goe to their places againe.

Sce: 5^{a}: Enter Belly:

Belly: oh that this Beefe and Brawne were but in jest
Soe might I sooner come vnto this feast.
Since they must bee eaten both, what need they striue
who shall bee first killd, or who last aliue?
The diuell's in the Cooke that cannot keepe
Them in the Kitchin: sure the old knaue's asleepe.
My stomack told mee some two howres agoe
'Twas supper time, but I scarce finde it soe.
They on whome I should eate (I meane the meate)
Are ready one another for to eate.
I'le waite on thẽ awhile, but when I light on thẽ,
Vpon the Table, oh how I will bite of them.
If I bee partiall, let mee bee hang'd vp
Tis not an oxe can conquer my vast gut.
Ile goe & whet my teeth for this good cheere;
In the meane time beware your Christmas beere.
Exit:

Brea: I told yee soe, lett's stay and see the ende [FOL. 110a]
Heere comes King Beefe: Cush: hee lookes not like Brawnes freind

Actus: 3^{s}: Sce: 1^{a}:
Enter K: Beefe with Sr Vinigar, & Sr Pepper.

K: Bee: What say our Knights? what doe our enemies meane?
To fight it out, or to forsake vs cleane?
Hast thou descri'de thier armie? if thou hast
Tell vs that wee may make the greater hast.

Vin: Brawnes sonne, Lord Souce comes dropping w^{t}h him to:
Oh wee shall haue a pittifull adoe.
Souce needes noe other weapon but his smell,
Twould choake one though hee were brought vp in hell:
Mustard is feareles, hee contemnes his foes,

300 *jest*] perhaps *jest.* 305 *asleepe*] ^{2}*e* altered from *a* 313 *Tis*] perhaps *'Tis*
318 *freind*] *in* minims indistinct 323 *thier*] *ie* altered from *ei* 325 *dropping*] *sic*
328 *Twould*] perhaps *T would*

Oh sayth hee, let mee haue them by the nose.
Thus my dread Leige, are you on each side layde for.

K: Bee: Courage my Subjects, what are yee affraid for?
Thinke you our forces can not maister theirs?
why what is Brawne? or what is Souce that dares
Compare his hoggish forces to our might?
Will they deny great Oxes sonne his right?
was not my father king of all the heard?
And shall his sonne by hoggs bee made affraid?
Haue wee not equall healpe? why then what care I?
Full flesh, stiff ribbs, & branches of Rosemary?

Pep: My Lord, your seruant vowes to stand
In your defence, as longe as heere's a hand:
And though I cannot wounde with deadly wronge,
Let mee alone to bite them by the tounge.

Vin: And vinigar sweares hee will play his part,
Though hee bee not soe stronge, hee's quick & tart.
Thou Pepper shallt inflame, and I will coole, [FOL. 110b]
Ile make them looke as though they were at stoole.

K: Bee: why thus it should bee. but what armes, what weapons
Shall wee aduance against these fearfull Capons.
I'le beare my Fathers horne to goare the slaues,
I there's a toole will send them to their graues.

Vin: what if your fathers horne cannot bee found?

K: Bee: Are there not horned beastes enough on grownd?
Or if the fields bee scant, & can not fitt yee,
yee may finde hornes to many in the citty.

Pep: Twas well thought on my Leige, shall wee bee packing?
And see that nothinge for the time bee lacking?

K: Bee: I come lett's goe; Brawne wee'l not doubt you S^{r},
Come vinigar. Vin: Ile clinge about you S^{r}.

Sce: 2^{a}:

Enter King Brawne, and the Lord Souce
his sonne, Must: his attendant to thẽ going
out:

K: Bra: Stand wide, giue roome, I come thou roasted slaue:
what dar'st thou thus on our owne grownd to braue?
Think'st thou that cause I'ue layne a longe time cold,
And in my wet hoale was allmost growne old,
Will giue my place to thee? if soe, take heed,
My power is comming on: Subjects proceed.
First heer's my Fathers tuske to meet thy horne,
It hath destroy'd full many a good Acorne.
Then heer's his bristles thrice as tough as stickes
Shall peerce thy sides farre worse then Butchers prickes.

K: Bee: wee feare thee not thou massy thick=skind Churle.
This horne shall thee from of the table hurle,
If thou put foote beefore mee; it shall goare thee
worse then the knife the Sire that went beefore thee.
Then heere's my rib, with this same cragged stumpe [FOL. 111a]
Thy sencelesse souce=sok't sides Ile thwack & thumpe.
Lastly these stubbed branches of Rosemary
Shall digge your hide, effaith Sr Ile not spare you.

K: Bra: Thou ouerrosted slaue, I feare thee not.
Take heede this crooked engine teare thee not.
Were't not I for some worthy followers waite,
My men & I would bid thee battell straight.
But stay till all come in, then black=burnt Beefe
Wee'l new beelard your ribbs, and to bee breife
Soe teare your bloud=dry'd grauy=swelling corse,
That hungry teeth shall rage with lesse remorse.

K: Beef: Well Brawne, wee'l meete i'th' shambles, 'tis not farre,
wee'l try of what solidity you are.

Exeunt:

Manent Vin: & Mustard.

Vinig: Thou villaine, why did'st thrust mee? think'st thou still
Wee are to gether in the Mustard mill?

374 *peerce*] perhaps *pierce* 396 *to gether*] perhaps *together*

Must: I thrust thee not thou musty sun=burnt slaue,
Yet thinkest thou the wall of mee shallt haue?
Proud Princkoxe know I am thy better nowe,
And waite vpon a worthier dish then thou.
Vin: O Traitour to my Lord King Beefe! haue I
Healpt for to moyst thy seede, when thou wast dry,
Haue I created thee, & gi'n thee spiritt,
That thou should'st striue my honour to inherit?
Vngratefull wretch, back or Ile wash thee thin:
Better thou hadd'st thy woodden house beene in;
Goe learne more wit of that graue ancient pot,
That will yet tell thee, if thou remembrest not,
How oft I haue encreasd thee at the bottome,
And more good turnes, allthough thou hast forgot 'em
Remember how I washt thy dreggs out once [Fol. 111b]
To serue i'th hall, that sau'd the Cooke a sconce.
Must: Do'st thou bely that honest fellow to?
O thou soure Varlet! what willt thou not doe?
That rayl'st against soe good a freind of thine?
Haue I not seene him with these eyes of mine,
When thou with butter should'st i'th' hall haue gon,
For thy more credit send thee in alone?
Vini: Nay if you goe to that, how oft haue I
Sau'd him, when hee hath beene at poynt to dye?
For when a loathsome rotten egge oft comes,
And the soft geare runnes all about his gummes,
Did not I Vinigar aneare him stande
To wash his mouth, hee would perish out of hand.
Must: Nay that's starke false, for though thou next him bee
I could driue out the sent, as well as thee.
Vini: But thou willt put him to a double paine,
H' had neede of bread to driue thee out againe.

405 *wretch*] altered from *wreth* 409 *encreasd*] perhaps *increasd*

Must: Thou sland'rest him & mee, think'st thou his nose
Can not indure such fleabitings as those?
That vs'd to choaking sauours, euer fighting
With the hot vapour of old Linge & Whiting.
Hee that with open nostrills lookes on these,
May spoone mee vp, and beare't away with ease.
But I'le indure noe more your scornefull taunts,
Try't with your bilbowe, let that seale your vaunts.

Vini: You're very hot Sr Mustard, 't would beseeme you
To know my presence, howsoe'r you [deed] deeme mee.
But since you are vpon the spur I'm for you.
And yet mee thinkes againe I'm somewhat sorry,
Knowing thou art mine owne begotten creature.
An[d] how I'ue healp'd to fabricate thy nature,
That nowe I should vndoe what I haue done
Does somewhat moue mee: yet 'twast thou beegun. [FOL. 112a]
Haue at your hide: alas! now in good trooth,
Say I should kill thee, I'de bee very loath:
But yet I'm vrg'd; ha, tell mee wilt thou yeald,
Beefore thou try'st the terrour of the field?

Must: yeald? think'st thou I was borne to shame
The honour of the high Sinapian name?
Or giue a foote of grownd to such as thee?
A poore thin rascall in respect of mee.
Noe try thy feeble forces; I'le not spare thee,
Come thou must dy, poore Vinigar prepare thee.

Vini: Nay Mustard stay, wee'l combate in the battaile,
Letts bleed in warre, & not in peace like cattle.

Must: It shall not serue your turne, or fight, or dy.

Vin: Oh stay I am not ready, too't by'nd by:
I'le but goe whet my sword, & come agen:
Lett's haue faire play, and equall ods like men.

442 *An[d]*] deletion may be a blot 459 *agen*] *e* blotted

Must: S^r try your blade one boute, 'tis keene enough,
Were but your selfe of answerable stuff.
Vini: Then giue mee leaue to take my fees, & runne,
And come vpon thee like a thund'ring gunne.
Thus I retire my selfe, now thinke on hell,
Thither Ile send thee anon, till then farewell. (Hee runns away

Sce: 3^a: Enter Belly:

Belly: Oh that this Beefe & Brawne would but haue donne:
I'm sure I pine for't. I that was a tonne
In compasse, now am lesse then any ferkin:
See but how much there wantes to fill this jerkin.
Oh how my gutts within my bulke doe rumble,
Sometimes they crawle, sometimes they rowle, and tumble.
All while this Cooke, this idle Cooke goes peaking [FOL. 112b]
For Beefe, & Brawne in euery corner sneaking.
Cooke and bee hang'd make hast. do'st meane to starue mee,
Send in the meate, or thou thy selfe shallt serue mee.
Ile not bee made to stand heere like a noddy,
Whil'st all my gutts runne vp and downe my boddy.
Nor yet? Ile fetch thee then, your chopping knife
Shall hardly from these teeth defend your life.
And yet it will, 'twere better to intreat him,
Then to prouoke his anger, and to threat him.
Therefore I'le goe in humble manner to him,
And to take vpp this quarrel I will woe him.
Till I perswade him to't Ile neuer linne,
Soe shall I stuff this doublet growne soe thin

Act: 4: Sce: 1^a: Exit to the Cooke.

Cooke: Oh who would bee a Cooke to sweat & swagger
For other men? oh that I had I dagger.
I'de tame this Beefe, and Brawne, & thier whole rout.
I meruaile how the Diuell this Brawne gat out,

469 *tonne*] *o* altered from *e* 488 *Exit . . . Cooke*] slightly raised 490 ^{2}*I*] *sic*

Or Beefe, when as I charg'd them goe noe farther
Then the stronge Barracadoes of my larder.
But yet they haue transgrest our sterne cōmand.
They're free, but I must to the danger stand.
And [I] though I stirre my stumps in what I can,
To giue content to all, yet eu'ry man
Will bee vpon my jacket; and the Cooke
Alone must to a world of busines looke
One hungry fellow runnes about & wishes
Hee could but finde the Cooke, I then the dishes
Should fly about. the rest of them they raue [FOL. 113a]
And cry where is our meate? this Cooke's a knaue.
And how alas can I prouide them meate
When as that's runne away, wch they should eate.
I heare that they fall out because I meant
To serue Brawne in first. but what's thier intent
After this quarrell that Ile search anon.
O that I knew but whither they are gonne.
I'de quicly fetch them in, & make them know
I am of some authority I trowe.

Sce: 2^a: Enter Belly:

Belly: Now is the time for him that coggs and flatters.
Now I'le stepp forth: Great King of dishes, platters,
Foule dripping panns & spitts: sole Lord, and Master
Of that good cheere, where of I'de faine bee taster.
Great Duke of Chimnies famous territory,
Heare mee with patience, whil'st I tell a story.
Cooke: Oh Mr Belly S^r I know your minde:
I'm sure you hope good victualls heere to finde.
I would I had it for you, but to say
The truth both Beefe & Brawne are runne away.
Belly. I know't (right W^rll) but I can tell you

507 *meant*] altered from *ment* 517 *where of*] perhaps *whereof* 521 *victualls*] *v* altered from *w*

Of all thier complotts; doe you thinke that Belly
Smells not out Beefe & Brawne in eu'ry corner,
I & Queene Mincepy, though some teeth had torne her.

Cooke. O blesse mee with this newes: Belly by Gis
I sweare thou shallt fare farre better for this.
Ile stuff thy greedy all deuouring panch
With delicates, wch shall thy hunger stanch.
Wee'l teach thee to forget to eate butterd fish, [FOL. 113b]
Some Phenix at the least shall bee thy dish.
O tell mee then but where as these extrauagants,
That wee to fetch them home may haue a gante.

Belly. Soe please your Sattin doublett's greasy grauity,
I'le tell you all I know. and now S^r haue at you.
Thou great Apollo, who's sweet harpe is strunge soe well,
Guide mee now speaking, that cann'st rule the tounge soe well
A bloudy quarrell must by mee describ'd bee,
But should they come, oh tell mee where to hide mee.

Cooke: Tut feare not man, I warrant thee th'art free
They'l tremble all, when they but looke on thee.

Belly. Then know (Great S^r) and I may tell it you,
They'r gonne to th' field to fight, beeleau't 'tis true,
You know the reason: & 't was my intent
To tell your worshipp what these roysters meant.
But I poore I by this meanes haue not din'd,
See how I'm falne away, see how I'm pin'd.
vnlesse your worshipp will vouchsafe [safe] some ayde
To mee now allmost desperate, I'm affraide,
T'will bee a meanes that I for want of meate
Shall bee my selfe made food for crowes to eate.

[Belly] Cooke. Feare not my noble Belgicus, for I
Will goe & look out for them by & by.
Know'st thou the place? Bel. I doe, for I heare say

539 *well*] binding may obscure final punctuation 554 *Cooke*] written below deletion

Tis at the Market, this the very day.
The houre's at hand. Coo: Then wee will goe & parte
That Beefe & Brawne, that beare's soe stout a hart.
Lett's hast, mee thinkes I heare the Scollers say
where is this Cooke, this knaue? hee's runne away.
Then followe mee my selfe will ende the strife [FOL. 114a]
with noe small weapon, with my chopping knife.

Belly. I ne're more willing went (the Gods bee thanked)
Then now, when eah stepp leades to a goodly banquet.

Sce: 3^{a}: Exit Belly, & Cooke.
Bread, Salt: Table: &c: stepp
forth againe, and speake.

Brea: How like you this my M^{rs}, I still thought
The matter would at lenght to this bee brought.

Cush: That Belly can as well bee hang'd, as keepe
Him selfe from Victualls, sure hee eates[s] in's sleepe.
His meate into his gutts hee doth soe ramme in,
That sure hee'd dye, should hee but heare of famine.

Salt. But to leaue this, I thinke it best to goe
Into some inner roome: Tab: And I thinke soe
For if perhapps in fury they come by,
Faith Bread, & Salt, & Trencher they must fly.
Or whatsoe're come next. Therfore letts all
Stay i'th next roome, still ready at thier call.

Act: 5: Sce: 1^{a}: Exunt omnes.
Enter at one dore King Beefe Queene Mincepy,
Sr Pepper, Sr Vinigar, with Spitts, & dripping
pannes, and such like weapons:

K: Bee: What not yet come King Brawne? then th'art a coward.
Speake good Sr Vinigar, of our foes hast thou heard?

Vini: I heard but now (my Lord) they were a cõming.

K: Bee: It may bee true, for hearke I heare a drũming.

565 *eah*] *sic* 572 *sleepe*] ^{2}*e* altered from *a* 581 *Exunt omnes*] *sic*; slightly raised

Sce: 2: [FOL. 114b]

Enter at the other dore King Brawne,
Lord Souce, Sr Mustard wth weapons as beefore:

K: Bee: What? art thou come? I see thou keep'st thy word.
Braw: yes Beefe & will maintaine it with my sword.
Descend from wordes to blowes vbraiding varlet,
Thou art my object, with thy Queene that Harlot.
Q: Mince: what? doth hee call mee Harlot? stand aside:
By Cocke & Pie, Ile make the rogue one eyde.
K: Bee: But peace faire Queene. But Brawne what do'st thou meane
To call our royall Queene, as bad as Queane?
K: Bra: Ile make thee know that thou hast wrought flat treason
To depose vs, when thou art out of season.
Vini: Well Brawne for once I will say thou ly'st,
But if thou speake soe once againe, thou dy'st.
Must: How now S^{r} vinigar, are you soe lusty?
Ile pay you now, allthough my blade bee rusty.
Vini: Thou know'st not how to handle it I feare
Ile deale ^{t}wh thee at cuffes: Mu: oh my eare. { they pull one another by the eares

Sce: 3:

Enter the Cooke with his Chopping
knife, with Belly:

Cooke. Shame take yee, that yee put mee to such feares
who hath set you together by the eares?
K: Bee: I now here's one will fight with vs [anon] I doubt mee
Cooke. Leaue of you vassailes, or I shall soe clout yee,
That I shall make yee all repent this trouble,
which yee haue put mee to treble & double.
Prostrate y^{r}sel[f]ues, & lay y^{r} weapons downe Belly ꝑtes them
Doe yee not feare your liues, when wee but frowne?
Omnes: Wee doe, wee doe, wee doe. [FOL. 115a]

597 *eyde*] *y* altered from *i* 598 *meane*] binding may obscure final punctuation 600 *treason*] perhaps *treeson* 607 ^{t}*wh*] *sic* 618 *Belly . . . them*] slightly raised

Belly. Gods blessing on your heart[e] good Mr Cooke.
Hee prouides meate, whil'st I but ouerlooke.
Cooke. Prostrate your selues yee whorsons at our feete,
Or else this chopping knife, & your neckes shall meete.
K: Bee: King Beefe, who was your Viceroy, resignes now
His crowne: and to your feete doth humbly bowe.
K: Bra: And I but now a King. Sou: & I the Kings sonne
Giue ouer to your mercy our Kingdome.
Q: Min: Good Mr Cooke for Gods sake pardon mee { Shee kneels
I am a silly woman as you see. { downe
Cooke. Rise vp, but Ile not brooke this bad vproare,
Beleiue mee Sirs but yee haue vext mee sore.
For wch thou Beefe in bonds of packthred tied,
Shall't first bee rosted, after broyl'd, & fried.
And Mrs Mincepy cause you made such hast,
To get forth, you shall bee inclosd in past,
With walls that shall surpasse these walls of mud,
Walls made of finest flower. Bel: Oh this is good.
Cooke. As for Brawne, Mustard, Vinigar, Pepper, you
With Souce, in your old mansions ay Ile mue.
Omnes. O spare good Mr Cooke this cruell labour.
Cooke. Cruell it is not, when you scorne our fauour.
But if you'l yeald to bee seru'd in by mee,
your punishmentes shall mitigated bee.
Omnes. Most willingly, most willingly. Coo: Then to tell yee,
The trueth, your cheifest guest to night is Belly.
Omnes. Oh woefull! Coo: Belly goe thou in & stay,
For preasantly Ile send the meate away.
Belly. Make hast for Gods sake: Exit Belly. [FOL. 115b]
Cooke. Brawne, th'art first in order.
Souce stand you backe, Ile keepe you in my Larder.
Mustard bee next thy Mr: giue attendance

630 *downe*] slightly raised 631 *vp*] *v* altered from *u* 636 *inclosd*] perhaps *inclos'd*

For if you do'n't, Ile make you with a vengeance.
Beefe come thou next. Bee: I should haue beene the first.

Cooke. What doe yee prate? well you shall fare the worst.
Pepper & vinigar, guard him well I pray,
For it is very like hee'l runne away.
you Mrs Mincepy shall alone bee sent. Exeunt omnes.
Soe get yee in. Souce Ile keepe you till Lent. Manent Cooke & Sou:

Sce: 4^{a}:

Cooke. The storme is past, this powerfull hand hath stilld
All troublous jarres: our Scene with peace is fill'd.
Noe scarre, nor breach our newborne quiet seauers:
If you'r pleas'd eyes, smile on our poore indeuours.
With you I come to make yet one peace more,
Not with my chopping knife, as heretofore;
But with this womans weapon to intreat,
Not to commande, as earst I did the meate.
For flowring stile, or phrase yee haue it not,
I'ue learn'd noe more then how to boyle a pot.
This for our labour: if the subject fit not,
As beeing to grosse, 'tis hard though it hit not
The time right home. noe matter how small reason,
y' 'aue rime, & though not good yet 'tis in season.
If you are plesd, loe heere a Sutor standes
wee neede noe Lawrell, crowne vs w^{t}h your hands.
But if your eares bee greau'd with such a toy,
y'are rid of vs, and soe clapp hands for joy.

Finis.

653 *do'n't*] *sic* 660 *Cooke & Sou:*] slightly raised 664 *you'r*] *sic*

Item 11 in J.a.1: Fols. 119a–133a ***Heteroclitanomalonomia***

Heteroclitanomalonomia is an academic grammar play in the tradition begun by the *Bellum Grammaticale* of the Italian Andrea Guarna. Leonard Hutton adapted Guarna's Latin prose into a Latin verse play, which was written by 1583 and acted before Queen Elizabeth at Christ Church, Oxford in September 1592.[1] *Heteroclitanomalonomia* seems to have been the first of the academic grammar plays written in English. The unknown author uses pentameter couplets, which he handles with considerable ease and sophistication. The play differs from Hutton's in a number of ways. Queen Oratio, who suffers from the war between Noun and Verb over precedence, is introduced in her own person, and, together with William Lily the grammarian, she persuades the two kings to make peace. Noun and Verb do not have parasites, but a separate subplot concerns Mr Ignorance, a schoolmaster, and the peddler Absurdo, an Autolycus-like figure with a pack full of 'all such wordes / As this whole Grammer=Land no where affordę' (ll. 375–6). Absurdo tries to sell Ignorance verb tenses, and finally Ignorance is attacked by the irregulars who, finding that he cannot furnish them with their missing parts, put a fool's cap on him. Whereas Hutton builds up to the battle between the followers of Noun and Verb, the war in *Heteroclitanomalonomia* is over in the first scene, and the remainder of the play is concerned with the difficulties of imposing peace on all concerned.

The most interesting aspect of the English play is its understanding of political tactics and complications. Hutton emphasized Participle's double nature and possible perfidy. *Heteroclitanomalonomia* shows the difficulty of controlling not only Participle, the traditional double-faced egotist 'for you both, but yet indeed for neyther' (l. 203)—but also the foot soldiers. These are the irregulars, the heteroclites, the verbs under Volo (who 'will not beare these wrongs'), and the nouns under Vis. Volo counsels insurrection: 'I'de you advìse / To take vp armes ageinst our carelesse kinge' (ll. 335–6), even though Fero asks, 'ffight wth our king? is that a thing so little?' (l. 338). The rebels become 'outlaws', and Volo articulates perfectly the position of an anarchist: 'I say we'ele ha' noe Lawes' (l. 463). Although Oratio feels sorry for them, and suggests building a spittle, the defectives complain of neglect. The disaffected former soldier is a stock type on the Jacobean stage.

The academic allegory exists on two levels. First, the play is presumably intended to explain Latin grammar. Parenthesis, the chorus, comments after Act III that the treasonous outlaws make the 'Grammer kindgome bleed', and give birth to Mr Ignorance as 'a token / Of their disturbance' (ll. 656–7). Ignorance

[1] See Frederick S. Boas, *University Drama in the Tudor Age* (Oxford, 1914), pp. 255–67.

sends his scholars to the 'Vniassitie'. In the final scene Lily, aided by the character 'Robin Robinson', confines the anomalaes to 'the grammer prison calld' *notandũ*' and the heteroclites to the dungeon '*Quæ gemas*' where 'Hexameters shall be there guard' (ll. 1122, 1130, 1133).

The identification of Robin Robinson is uncertain. Bowers (p. 122) was confident that he was Thomas Robertson, Dean of Durham, 'who added the section on heteroclites to Lily's *Latin Grammar*', but David L. Russell proposed Robert Robinson, a Londoner whose *Art of Pronuntiation* was published in 1617 (STC 21122). Russell claims that the characters share Robinson's 'interest in correctness and maintaining standards'.[1] Hilton Kelliher, who believes that a mistake of Robin for Thomas would not have gone undetected, proposes another Robert Robinson 'who in 1607 matriculated from Trinity College, Cambridge, and whom the *Alumni Cantabrigienses* identifies as a preacher at Gedney, Lincolnshire, in 1614 and as a schoolmaster'.[2] Nevertheless, Thomas Robertson's work on heteroclites remains entirely relevant to the play, and (despite the change of name) he is still the most likely person to be figured as the character Robinson.

The play also represents the college environment in allegorical terms. The outlaws are looking for those 'that care not / what they Doe speake or doe . . . Newcomes of the Colledge' (ll. 442–5). The Epilogue informs the audience that this story is 'true, but coverd' with an Allegorie' in which the actors 'ment / By the Defectives freshmen to p̵sent, / wch daylie like irregulers rebell / Ageinst vs seniors' (ll. 1152–6).

The reference to 'vs seniors' is one of several titbits of information about the production. The purpose of the performance is 'to rub vp our grammer endes' (l. 12). It may be presumed that at least the chief characters were played by seniors; whether the outlaws were played by freshmen it is impossible to tell. The chorus is Parenthesis, who calls for music between the acts: 'Pray lett them twange their instrumentȩ a little / Till I am tyrd'e [i.e. dressed], for I play Participle' (ll. 171–2). There may have been other doubling of parts as well. The presence of music and the five-act structure suggest a formal performance, but no particular scenery or special equipment is necessary. There must be some way of representing 'within' (IV.ii), and a door for Robinson to knock on in IV.iii. Props are confined to letters, Absurdo's pack, and whatever was necessary to suggest the war injuries of the defectives.

The manuscript of *Heteroclitanomalonomia* measures 150 × 193 mm. There are no ruled margins or running heads, but there is an ample margin, 25 mm to the

[1] David L. Russell, p. 27 (edition cited on p. 3, n. 7).

[2] Private communication to the editors. See John Venn and J. A. Venn, compilers, *Alumni Cantabrigienses*, Part 1, 4 vols. (Cambridge, 1922–7), iii. 473.

speech prefix on the left, and almost 33 mm on the right. The hand is fine and regular: secretary except for the use of italic for stage directions, Latin, and the letters which characters send to each other. The rather old-fashioned look of the hand comes partly from the use of long descenders. Other characteristics of this hand are extensive abbreviations (the hand has different symbols for *pro*, *per*, and *pre*), reversed question marks, double hyphens, and a flourished *a*. The hand has no majuscule italic *V*; the scribe uses lower case consistently in the speech prefixes (e.g. *volo*, *verbum*) and once, in l. 688, writes '*Verbũ*' in italic with a secretary *V*.

The scribe was fussy and made many small corrections as he went along. When he thought two words were too close together, he drew a line between them (e.g. l. 833, 'thoushallt', l. 966 'alittle'). It is difficult to distinguish certain letter shapes, especially *D*/*d*, *Y*/*y*, *W*/*w*, and apostrophes are placed carelessly, often over letters (e.g. in l. 903, 'Ile' has an apostrophe over the *l*; the word is rendered in this edition as 'I'le'). Occasionally a question mark appears together with a comma or period. The decorative title, *Heteroclitanomalonomia*, is written in black-letter, which appears nowhere else in J.a.1.

The title, like the generally elegant form of the manuscript, makes it clear that this is a fair copy, perhaps written for presentation. Nevertheless, the adequate stage directions suggest that the present manuscript may have been copied from another, either used or intended for performance: '*Goes to the dore|to aske whose' there|then returnes*' (ll. 848–50). Somewhat puzzling in this respect is the treatment of two passages, and gaps in the text, towards the end of the play. In the first (ll. 1079–82), Robinson turns to the Heteroclits and says:

> To ꝓceed Wth these Hetroclitę heere ar 3 articles framed
> ageinst them.
> * * * * * * * * * * * * *
> what say you guilty or not guiltie.

It is possible that he merely waves the paper without reading the charges at all, but this explanation seems less satisfactory for the passage in which Verbum tries to right Priscian's wrongs (ll. 1105–11):

> *ver*: Come Priscian we doe now accept thy writing,
> Thou art misvsd, we'ele labour for thy righting.
> Read these his supplications Lilly.
> *Lillie*. * * * * * * * * * * * * *.
> *ver*: what thou requestest heare is nought but reason,
> Those men w'eele bannish, & their wares w'eele sease on.
> *Lill*: * * * * * * * * * * * * * * *.

In this case it seems more likely that the scribe thought the passage would be supplied from another source—perhaps an extract from Lily's *Grammar*. The speech prefixes for Lily must otherwise remain unexplained.

At some time a reader (W) who was having difficulty with the secretary hand annotated several of the words incorrectly. On Fol. 122b, for example, '*taillffs*' is written above 'taylesse'; '*shanie*' is in the margin as a note to 'shame', which is underlined; and the word 'Raskall', also underlined, is transcribed in the margin as '*skillfull*' (see ll. 279–81). The correction attempted at the beginning of the Chorus to Act II appears to be in the ink of Hand W. Where the text originally read, 'Now I presume ageine Parenthesis' (l. 468), the *p* in 'presume' has been mistakenly struck through, thus enabling Parenthesis to 'resume' again. W attempts a few similar corrections, adding a question mark to line 282 and regularizing the metre of line 569 by adding 'an' to 'others'. W's handiwork does not appear again after the third act.

1613. [Fol. 119a]

Heteroclitanomalonomia:

Prologus./

Wee purpose to ꝑsent vpon o^{r} stage
A Battaile wch was fought before o^{r} age:
Nor ieaste nor earnest is our whole intent,
But as you'le take it, soe we'ele say t'was ment.
The Grammer kingȩ both greedie of commaunde
Each in the cheefest place of speech would stand:
Till *Lillie* settȩ betweene them both a barr
And wholie pleades for peace, thus endȩ the iarr,
Our cheefest aime's at that wch followe's after
Among the maymed, wch may move some laughter
Wee doe it, to rub vp our grammer endes
A trifle, yet I hope t'will please our freindȩ.

0.1 *1613.*] added in hand X (see above, p. 13) 0.2 *Heteroclitanomalonomia*] written in black-letter 4 *intent*] *i* altered from *e* 9 *peace*] ^{1}*e* has flourish of *a* 10 *aime's*] *a* blotted, ? altered

Act^{9} j^{9} Sce: j^{9}

After an alarme, & some other signes of Battaile &c. Enter Priscian wth a Broken=head wch he had taken in the warrs betweene *Nomen & verbũ*.

Priscian. Cease, cease you warlike instrumentꝭ of Battaile
Ly still yee weapons of Grammarian⟨s⟩ Mettle
Nomen, good Verbum, yf you be a Christian
Leave armes, abuse not poore oppressed Priscian.
Laugh not my Masters, for this is noe dalliance
The nowne & verbe [be] are fallen out at variance
I then my fellowes being somewhat boulder
Beare away this, & somewhat on my shoulder.
Nor onlie I, but manie more (God wott)
In these dread warrs defectꝭ & woundꝭ have gott.
In verie sooth t'would greive your stonie hartes
To see [w] how hott each armie play their ꝑtes
ffirst both the kingꝭ doe gather vp their *Numbers* [FOL. 119b]
And each their Captaines from their Cowardꝭ sunders.
The articles as *Hic*, *hæc*, *hoc* and others
Begin[e] the Battaile sparing not their Brothers.
To them three *Parsons* of King verbum's side
As *ego*, *tu*, and *ille* soone replied.
Then the declensions in their severall rankes
wth Coniugations 'gins to play their prankes.
The Gerundꝭ & the Supines they stand by
To drowne wth Drumbes & ffifes the maymed's cry;
Verbes wch have lost: their limbes, *quàm plurima dantur*,
Et quæ deficiunt genere adiectiva notantur.
Heare lies a ꝑfect=tence of Verbum's troupe,

19 *warlike*] *l* altered from *k* 20 *weapons*] *p* altered from *w* *Grammarian⟨s⟩*] *s* obscured by blot 34 *Begin[e]*] cancellation may be minim of another *n* 35 *side*] *e* altered from *i* 38 *Coniugations*] *Coniug* possibly italic

And on the other side yf you would stoope,
You might take vp a hundred thousand Cases;
There standeth ffero making crabbed faces
ffor his lost ꝑfect=tence, an other lies
Lame of his leggꝭ. *Cæcus* he wantꝭ his eies.
Volo crowdes valiantlie into the presse
But strayte (alas) returns imperativelesse.
Dice wth speeches thinking to prevaile
Comes back poore *Dic* cut shorter by the taile.
Soe Vis the nowne [th] although he thought no harme
Yet in this skirmish lost his giving arme.
Fumus he smoak't fort', what with shott & thunder,
He never since could find his plurall number.
The Guns so bounc't in *Quidam's* eares & face,
Hee's thick of hearing in his *Calling* case.
what more? some verbes grew hoarse that wth their noyse
Some lost their [at] *Active* some their passive voyce.
In these hott warres, where neyther Verbe nor nowne
Saw their owne fellowes groveling on the ground,
The Adiectives wth their substantives did buffett,
And I wth ꝑting them was soundlie cuffed.
An alarme. See heere the signe; But harke, *Gerundes in Dum*
Doe sound their Trumpettꝭ, Supins beate their Drumbe.
O fie vpon these warres, & fowle ambition
Each ageinst other, both ageinst poore Priscian.
This is no time noe place for me to stay.
The nownes & Verbes draw neere, I must away.

Exit.

Actꝰ Sce: 2. [FOL. 120a]

Enter Oratio held by Nomen on one
arme & by Verbũ on the other and
after them *Lillie*.

50 *imperativelesse*] minim obscured in *m* 58 *thick*] *c* altered from *k* 59 *hoarse*] *a* interlined over caret 63 *substantives*] ^{2}*s* altered from *t* 72 Sce:] *c* altered from *e*

verb: *Oratio* is mine, my handes the first did light on her,
Nomen thou wert' not best to cast one sight on her.
Nom: Peace *Verbũ*, peace, thou nothing art but worde
Renowned *Nowne* shall gett her wth his sworde.
Orat. Houlde both yor handꝭ. *Oratio* speakes in teares,
your civill Broyles her cheekes wth bloud besmeares.
ffor be you sure, that such a death of Wordes
As this your battaile every howre affordꝭ
Must needꝭ hurte me. the world was never better
Then when we did enioy our [sev] ser⟨v⟩ant *Cæter*.
whome ye have slaine, wth manie of his order,
All wch ar dead; whatꝭ this but open Murther?
yf you goe forward as you have begun,
All *Gramm*er=Land will quicklie be vndon.
I feele that Barbarisme nighe approches,
when our best suƀtes all doe halte on crotches.
Some loose their [bo] legges, their armes yea bodies, and some
Withall do loose their lives, thinke you this hansome.
Decree a peace; of you this onlie seeke I
Præstat enim regredi, quam mala cæpta sequi.
Nom: Il'e pawne my creddit, yf thou wilte beleeve me.
Il'e doe what eare I may for to releeve thee.
But Il'e not loose my precious [prise] prize, thatꝭ flatt;
What's mine is mine, & I will have but that.
verbũ. whatꝭ thine proud nowne, base, lowsie Beggers peasant,
I would I could but once espie thee sease on't.
Orat: Leave of these Lowsie [p] Beggars yf you love me,
The very name of Peasant much doth move me.
Lillie Since she is his & yours, but wholie neyther,
T'is best you both should for a respite leave her,

77 *Nomen*] *e* altered from *o* 80 *handꝭ*.] . may be tail of *h* 84 *hurte*] *r* interlined above caret 85 *ser⟨v⟩ant*] *v* blotted, ? *w* 86 *slaine*] *l* altered from *t* 89 *quicklie*] *k* altered from *c* *vndon*] *o* altered 92 *armes yea*] interlined over caret 93 *lives*] *v* altered 96 *wilte*] *t* altered, ? from *l*

Or both possesse her but in quiett manner,
you doe not well, thus for to hange vpon her.
please you to le[a]nd yor eares, you strait shall see,
Some articles to wch both shall agree.

Nom: Content saie I not. that I feare to combat
wth that same Verbũ, whom If I can come at,
Ile teach him [w] how to meddle wth my Wenches
he hath enough to doe wth moodę & tences.

ver: And I agree for this my Mistres love [Fol. 120b]
Not yours, an other time these wordę shal ꝓve.

Lillie. Why now you deale like Kingę. to end this t[〈 . 〉]oyle,
Be pleased therefore for to sitt a while,
Oratio in the middle like a Mother;
Nomen on one side, *Verbũ* on the other.

Ora: Speake then (grave Lillie) as a wise man should doe,
where be these *compositions*, say what would yee?

Lillie. *For the due ioyning of wordes in construction it is to be vnder=standed, that in Latine Speech there be three concordes, the first betweene the nominative case and the verbe, the sea=cond betweene the substantive and the adiective, the third be=tween the antecedent and the relative.*

Ora: why? this is true, but now we aske of thee
How may we make these enemies agree.

Lillie. The 2 last Concordę doe agree alreadie,
Although the nowne & verbe be somewhat headie,
About thee Queene Oratio, as it seemes:
wherefore in this my iudgm^{t} fittest deemes,
That yf yee doe consider Nomens case,
Why, then the Verbe shall take the cheefest place.
But yf yee doe respect the *Persons*, then
I take the nownes to be the better men.
Thus both shall Raigne but at their severall seasons

137 *severall*] *v* altered from *a*

In an Oration for thaforesaid reasons.
How likes *Oratio* this [t] devise for peace?
Orat: ffull well (good *Lillie*) doe they Nomen please?
Nom: They doe, & doubtles he were verie crosse
That would not lett wise Lillies sentence passe.
How please they Verbũ? *ver*: all as well as may be,
I neer' sawe better articles then they be.
Be you but pleasd' in publick wee'le reveale them,
And with our hand, & privie Signett seale them.
Ora. we all consent; Lillie lead on the way
we meane to aske the next schoole leave to play.
ver: which being done, dread Queene) I will returne,
To comfort those wch in these warres doe morne
And send backe trustie Participles armie,
wch he eare long will have ꝑvided for me./

Exeunt omnes./

Chorus. [FOL. 121a]

Wee'le be as neat, as you know who before vs,
I'le tell you Sirs, our play shall have a *Chorus*.
And I am he, I speake it with an Emphasis,
My hoopes expound my name it is Parenthesis.
The eplilouge, & prolouge they doe stand,
Like these same Semicircles in my hand.
wch thus encloses wth their Semimoone,
Our needeles sportȩ, wch well may be vndon.
But to my part, the kingȩ as you have seent',
Are freindȩ, the *Compositions* are in print.
But now the nownes & verbes are fallen in,
Anomalaes & Hetroclites begin.
Sitt heere a while, & you shall see some sport,
wch they shall make, then laugh, & thanke vs for't.
Yet stay a little, lett me not forgett you,

144 *sawe*] *w* altered from *y* 149 *done*] ? *Done* *Queene)*] opening parenthesis omitted
159 *eplilouge*] *sic* 162 *needeles*] *d* begun as *l* 163 *But*] *B* altered

One thing I have, for wch I must entreat you
Pray lett them twange their instrumentꝭ a little
Till I am tyrd'e, for I play Participle./

Act^{9} 2^{d9}. Sce: j.

Enter Participiũ Sot̄ wth two lr̃es.

Par: The *Grammer* fallen at oddꝭ, this newes is sad,
I feare wth ꝑticiple t'will goe bad. *Legit lr̃as./*

To his right trusty subiect Don Pedro
Participio inhabiting in Accedence Alliẽ
beyond great Possum, neare to the
abode of Signior Eo and Queo.

I know thou art not ignorant honourable Participle with what insolence, and hawtiness of minde Poeta king of the Nownes hath made an insurrection, and how rashlie he præsums ageinst the law of verbes to vsurpe the supremacie in an Oration, in so=much that wee are compelled to take vp armes though very vnwillinglie ageinst him to beate down his pride and save o^{r} owne honour, wherfore being you know full well how much ye are beholden to vs, as for your tence and signi=fication you should performe an action befitting your fidelity yf you would præserve the common opinion we have allwaies had of you from the beginning. by adioyning your bandes, and your selfe to ours not [ol] only for the præservation of vs, but also for your owne safetie, of which you should have cause to despaire should our empire be once overthrowne farwell, and hasten thy coming as thou art able.

Amo verbũ.

Stand toot' stout Nownes, much honour may
you merrit [Fol. 121b]
To gett Oratio & proud Verbũ ferritt.
ffight on stout verbũ, noe lesse mayst thou win,
To gett oratio, & beat [Verb] Nomen in.

188 *beholden*] *h* has tail of *y* or *f* 192 *[ol]*] *l* may be beginning of *f* *also*] *l* altered from *s*
199 *lesse*] ^{1}*s* altered from ? *e*

Strike on both ꝑties, heer's none will forbid it,
Brave be mine honour. this is Grammers credditt.
I'me for you both, but yet indeed for neyther,
Though I love you, beleeve me I had rather
Plesure my selfe then both. Then Verbe & Nomen
Learne this of me, ti's best for you to know men,
before you trust them, t'is a common Proverbe:
Doest thinke, that I'le neglect my ꝓfitt? no verbe.
They write to me to ioyne [te] my Bandꝭ to theirs
And each of them the losse of Kingdome feares;
I'me glad of that; when two extreames do strive,
Then is the middle very like to thrive.
And yet because they shall not know my meaningꝭ,
A few Il'e send them of my courser gleaningꝭ,
The verbes ꝑhappes a few Neutro passivaes,
Lest they mistrust me. (oh the'ire ꝑillous slye knaves)
The nownes some Toyes wch end in Tor & *Trix*.
ffor such slight thingꝭ noe politician stickes.
I have enough to serve my turne & more,
why, I can number halfe an hundred score,
As those in *Ans*, *Ens*, *Tus*, *Sus*, *Xus*, *Rus*, and *Dus*
My souldiers doe in number passe the Sandust.
And more [proud] then that, proud Nomen do but thwart y^u^ vs
We'ele rayse the spiritt of our long dead mortuus.
Well, well, I will be king at least, beshrow me,
But neyther nowne nor Verbe can keepe it fro me.

Enter verbũ, volo, fero, Dic.

Act: 2. sce: 2.

ver: Since wars be donne, lett eu͛y of our nation,
Betake him to his prop[p]er habitation.
Par: How now king *verbũ*, hah? what fled? now wellaway,
How goes the warrs, hath nomen bare the Bell away.

216 *the'ire*] *sic* 220 *score*] *c* altered from *k* 232 *bare*] *b* altered from ? *g*

ver: why they be donne; I scarce had seald' my letter,
When Lillie ioynd vs freindę. *Par*: Better, & better.
Thus while I make my selfe of all soe sure, *aside*.
I'me guld'e of all; who can such fatę endure?
But ar the Warrs brake vp? Im'e glad, beleeve me,
ffor both yor sakes, that onlie doth releive me, [Fol. 122a]
Had it not binne for you, the Participle
Voyde of all helpe, had still remaind' a Cripple.
yf you had chanct' to loose yor domination;
where should I had tence or signification?
Had Nowne vntimelie fell in that dissention,
what should I done for gender, case, declension?
If both had died, my dammage had binne bigger,
where should I had my Number [of] or my figure?
All wch considered, sure he were not worthie,
Such kindenes, who would not be carefull for the.
ver: I know you're carefull, & I thanke you for't'
Although your kindnes at this time came short,
I must impute it to my too late sending
And not in anie wise to yor offending.
But I am glad you did a while with=hold yours,
ffor had I had a verie few more souldiers,
All Nomens troupes had binne of life bereft,
There had not then one Adiective binne left.
Par: Me thinkę, that you are reasonably tamed
As far as I see, manie of yours are maymed.
ver: T'is true, t'was our hard fortune, & o^{r} fate
That thus massacred them; But say, relate,
How eũy wound to each of you befell,
And by what mischeefe? *Dic*. ffaith I cannott tell,
But as I'me Dic, I in the wars could never thrive
ffor there I lost the tayle of my Imꝑative.

238 *doth*] *d* altered from *b* 240 *had*] *d* altered 254 *verie*] *v* altered 255 *binne*] missing one minim 258 *maymed*] *e* blotted

I, as my vse was, when I see one come,
Turning my Back, a Nowne strooke of my Bum.
volo: And I poore Volo being somewhat willfull,
was though I say't my selfe not quite vnskillfull
And yet my iackett some Nowne had a pull at,
My Passive voyce was struck of wth a Bullett.
Fero. But silly ffero bore the brunt of all,
There did my ꝑfect=tence vntimelie fall.
But yet I hope your Lordshipꝭ war=munition
Will for our labours give vs restitution.
Oh to be ꝑfect is an happenesse,
And to be tenced is to me no lesse:
Then that I may be ꝑfected, & tenced,
Let me (I pray) with your rewardꝭ be fenced./
Dic. But to goe taylesse is a worse disgrace, [FOL. 122b]
ffor shame I dare not looke you in the face,
Ô that same Raskall cald *Apocope*,
Could not he ban'ge me, but must take my E.
Wherfore my Lord, my elboe daylie itches,
Hoping you will restore my long lost breeches.
volo: My dammages (dread King) you knew before,
Denie not then my losses to restore.
ver: How now (my freindꝭ) for recompences sute yee?
I hope that you have donne, but whatꝭ yor dutie.
yet in Rewarding woondꝭ I would not lagger,
But that these warrs have made me quite a bagger.
Wherfore be aunswerd', nay the matter scan not
ffor I am forc'd indeed to say I can=nott.
Seeke for rewardꝭ? now all my moodꝭ defend me,
I'me allmost mad; Good Participle 'tende me./

267 volo:] below line 270 *Passive*] *i* altered from *a* 279 *taillffs* written in italics above *taylesse* in hand W 280 *shame*] underlined in hand W; glossed as *shanie* in margin 281 *skillfull* in margin in hand W in italics *Raskall*] underlined in hand W 282 *?* added after *E.* in hand W 287 *sue* written in italics above *sute* in hand W 288 *have*] *v* altered from *t* 292 *can=not*] = uncertain 293 *rewardꝭ?*] *?* altered from ,

Act : 2 sc: 4. *Exeunt verbũ & particip:*

Dic. Is this a recompence for all our warres?
Is this a salve fit for soe daungerous scarrs?
I'le tell you, Sirs, although I make noe grudging,
Well may you thinke, I take the same in Dudgen.
T'hath stird my collor much, & can you blame me?
Whie; losse of Tayle it cannot chuse but shame me.
volo. Indeed thatȩ much, but my commaunding moode
His taile, head, bodie, all [all] ar drencht in bloud.
How may I storme for this same loss of mine,
wch in the warrs did all commaund resigne?
But verbũ I will make thee know (thou Base theife)
I will not beare these wrongȩ. I have noe P⟨a⟩assive.
Fero. High hoe, I'me active passive both together
Yet to say truth, I ꝑfect am in neyther.
I want that tence, yet simple though that I am,
I'le goe to Verbum, once againe I'le try him,
yf he will make me restitution for it,
If he denie it, I must needȩ goe borrow '[i]t.
Dic. ffy, fy, noe ffero, t'is a shame to borrow,
I'le make him give the one before to morrow:
Be ruld by me, we'ele trounce that thanklesse Verbũ,
I'le tell you Sirs, I'le make this same a deare Bum.
Hee'd better brake his neck from of a Steeple,
Then vsd' vs thus, ti's that same Participle, [FOL. 123a]
wch like a Hangʒon or a fflattring Parr[i]asite,
Gettȩ all from vs, he'ele still be sure to tarry by't.
Fero. Hee's tarried by it so long, that now the [Horson] whorsone
Hath gott into his fingers every Person;
My moode Imꝑative it wantȩ the first,

297 *scarrs*] accented *c*, ? altered from *a* 303 *[all]*] ²*l* partially obliterated 307 *P⟨a⟩assive*] ¹*a* blotted, ²*a* inserted over blot 320 *Parr[i]asite*] *a* interlined above deletion 322 *tarried*] *e* interlined above caret; *tarr* underlined in hand W 323 *tarried* in margin in hand W in italics

ffor wch to ask verbe yet I never durst.
Because I knew he'd[e] given him so manie,
That now for me to spare he had not anie.
Well well I knew the time, when t'was a fashion,
That eṽy nowne of whatsoever nation
Should be the third, now they ar not affeard,
To be the first, the seacond, or the third.
Nay more; this was confirmd by ꝓclamation,
Cryd' in the Market=place by Evocation.
volo: This must be lookt' to, ffellowes lettę be wise,
yf volo might be heard, I'de you advise
To take vp armes ageinst our carelesse kinge,
We'ele make him know, that Beetles have a stinge.
Fero. ffight wth our king? is that a thing so little?
Ile tell the volo, that I am no beetle:
Nor will I take vp armes ageinst our King,
So I my selfe may to worse mischeefe bring.
As I am ffero, I have borne some losses,
And ear' I'le do't, will beare farr greater crosses.
Besidę whereas King Verbum did denie me,
Perhaps he did it, onlie for to trie me.
Had I my ꝑfect=tence, then thou should'st see,
My noble Volo I would be for thee.
Dic: But give me leave (good ffero) I beseech thee,
In this case I am able for to teach thee.
Think'st thou to gett thy tence againe? (o wissard)
Take this from me, to gett from verbũ t'is hard.
wert' in my case, thou should'st not go to ask a Tence,
Wee'd quicklie [fett] ferritt Verbum from the accedence.
wer't in my case, quoth I,? why I am worse
I have more reason *verbũ* for to curse.
But heer's an Anvill, wch yf I can hammer,
I'le quicklie set on fire Will *Lillies Grammer*.

337 *Beetles*] dot over ²*e* 354 *I, ?*] *sic*

volo: I am resolvd'. yf fortune d'oe but favour vs, [FOL. 123b]
Proud *verbem* ear' long shall leave to triumph over vs.
Let vs turne out=Lawes, & who ear' we meete,
We'ele rifle [him] them, be it in open streete.
Fero: why then I see I must beare what betide me,
I'le follow them; Ill ffortune fall beside me.

Exeunt omnes.

Actus 2 *sce: 5.*

Enter Absurdo laughing.

ffaith, Don Absurdo, th'art a noble Lad,
Thy wares will vtter, be they nee're so bad.
But shall I say, what heere I have to sell?
Let me put downe my pack, & then Ile tell.
It was not long since I did vnderstand
Of the great difference in Grammer=Land
Betwene the Nowne & Verbe, vpon wch muttring
I did presume my wares would want no vttering.
I laded then an Asse wth all such wordes
As this whole Grammer=Land no where affordę;
wth out worne phrases, & ould coniugations;
Now some [new] young startvps, following still new fashions
would needę se all my wares, each for them seekę
Nay more; they faine would learne Absurdoes trickę;
And speake his speech, but marke what follow'd after,
Pray vnderstand me, for t'is worth the laughter.
An olde, grey=[headed] bearded bound=head father came,
I thinke they cald' him Priscian, yea, the same.
you may not speake (quoth he) such Barbarous phrases,
A word ill spoken oft great tumultę rayses.
No sooner spake he, but a moodie Squire
Amongst the companie sett all on fire
wth these his speeches, lent him such a Clap,

358 *d'oe*] *sic* 367 *ffaith*] *th* altered from ? *d* 368 *Thy*] *h* altered from ? *c* 381 *speake*] *p* altered or blotted 384 *yea*] *ea* written over in the ink of hand W

As made the bloud come trickling downe his cap.
Absurd, Absurd cryde he, now others followed,
Crying & pelting him, I whoopt' & hallowed,
My Asse he kikt', my wares were all vntied,
ffresh companie came, Absurd the ould man cry'de.
you hurt me, you shall awnswer't ere be long sir,
I hurt ye not quoth I, you doe me wrong sir.
But then thought I, t'is best for me to leave them [FOL. 124a]
Soe tooke this pack, & lest I should deceave them
That were my Customers, there I left my Asse
Glad that I scaped soe, & thus did passe;
As for the Beast, my customers may ride him,
Soe for my scattred wares lett them de[s]vide 'm.
This only I reserv'd, wch yf I cry it
I know that manie ḣeere about will buy it.
 who wantę a number, case, or gender
 Of anie nowne I can it render.
 who ever lackę a maimde verbes moode
 Come vnto me I'le make it good.
 Come along ye Laddę see what yee lack
 And ease absurdo of his pack.
But stay (Absurd) heere Customers are none,
All civell are, thy harborers are gone,
Sure, sure I mist my way, I may not tarry heere,
I must to Ignorance, I am his Carrier.
But soft, who are those fellowes that are lurking
Each of them beares some writing on his [⟨.⟩] ierkin.

Act: 2 sc:6 *Enter to Absurdo volo, fero, dic.*

Fero. Say ffellowes, han't we gott vs pretty prizes?
volo: yea. lett'ę take heede, lest that the Country rises,
ffor we have made shrowd vprores. *Fero.* ô tis best,

393 *kikt'*] ^{2}*k* altered from *c* 398 *deceave*] *d* altered from *b* 407 *maimde*] *d* altered from *b*; underlined in hand W *moode*] *e* faded, possibly a comma *maim'd* in italics in right margin in hand W 410 *absurdo*] underlined in hand W

Till we be ꝑfect, few shall take their rest,
Poore Tulo he lies breathles, voyd of sences,
But I am glad I ha' gott his ꝑfectences;
Thus have we somewhat. *Dic*. yea, you may be glad,
But honest Dic amongst you fareth bad,
I am a curtall still. *volo*: soft who comes hither?
Ab: A foe or freind, to out=lawes, choose ye whether,
My name's Absurdo famous through most nations,
ffor bringing vp strange orders, & new fashions.
Dic. wherefore cam'st thou to this place? *Abs*: don't yee see?
Only because the place came not to me.
vol: welcome in faith (good Don.) but pree'the tell vs.
what busines ha'st? *Abs*: I see you're merry fellowes;
your letters termes yee Out=lawes, your ar well mett,
I'le tell you what; poore Priscian wantꝭ a helmett,
what betweene you, & me & such as we be
All out=lawes, all vnrulie, nee're will he be [FOL. 124b]
without the Red=worme crawling downe his pate;
Soe I with other youngsters vsd him late,
And hither fled I. *Fero*. Ô but heer's no tarrying,
All heare are freindꝭ to Priscian, nothing varying
ffrom his good orders; Some there are that care not
what they Doe speake or doe, seeke them, & feare not,
As country Lobbes, & clounes, that have noe knowledge,
Rude Pædagogꝭ, & Newcomes of the Colledge.
The'ire thy best chapmen; we must hence away.
Ab: Tell me before you goe where did you lay,
Dead Tuloes furniture, I should vse his moodes:
Fero. Those you shall find heere by in Errours woodꝭ.

424 *you*] cancelled mark above *y*; ? penslip or start of *s* 425 *fareth*] underlined in hand W; *fareth* in margin in hand W in italics 426 *curtall*] *ai* written over *a* in hand W, partly obscuring ^{1}*l* 427 *out=lawes,*] comma uncertain, attached to tail of *f* in *soft* (l. 426) 433 Abs:] *A* begun as *I* 434 *letters termes . . . your ar*] *sic* 441 *nothing varying*] ^{1}*g* blotted, *v* extends back into ^{1}*g*

what we have left of his (Absurdo) take it.
we did ꝓvide our selves of what we lacked.

Ab: Thankes (good Anomalaes) I must goe seeke them,
Tulo Tulebam? who can choose but like them?
ffarwell vnto you all. *volo*. the like to you.

Dic. Well, well my Masters, heer's adeale a doe
ffor moodes & tences & such other triftles,
But I poore dic must goe away wth nifles
But now what to be donne? you have yor tooles.

vol: Lett'ꝭ goe about to all the Country schooles,
Set them a gag to break that head of Priscians,
They have noe knowledge of yor new additions.
Stand too't stout Brothers, since we are anomalaes,
Noe rule shall governe vs, I say we'ele ha' noe Lawes.

Fero: very good counsell, since we are anomalaes,
I say wee'le goe at Randum, we will ha' noe Lawes.

Exeunt omnes.

Chorus

Now I presume ageine Parenthesis,
Come for to tell you what is ment by this
wch in the second act [I] you saw before,
In livelie manner acted; & noe more.
ffirst then the verbes would have their losse restord
Having repulse they thretten fire & sourde,
These verbes be ꝑillous Rouges, for in a spleene
They kill & slay, the like was never seene.
ffero meetꝭ Tulo voyd of all defence
Knockꝭ him downe dead & steales his ꝑfect=tence.
Noe lesse doth volo to some Verbe or other
And takes vp Vis & velle, but their Brother [Fol. 125a]
Poore Dic I meane, gettꝭ nought at all amonge them
Thus strive they to avenge him that doth wrong them:

452 *Anomalaes*] *o* altered from *a* 462 *stout*] *ou* altered 468 *presume*] *p* deleted by hand W

But as they goe, they meet wth Don-Absurd,
They tell him of their prize, who at a worde,
Goes backe, & findę the ffurniture by chaunce,
And selles them to his chapman Ignorance.
Thinke that the Nownes doe play like Reakę wthin,
Sitt heere & see, now Doe the game begin.
yet for a while, lett Musick make you merry
To ease our actors wch be almost wery.

Act^{9} 3^{9}. sc: 3.

Enter [*now*] *vis, fumus, Aliquis.*

vis. Nomen for peace entreatę, from war dehortę,
Thretning dred punnishment to vs (consortę)
If that we treat but wth our lookes of fight,
soe direfull is it in his kinglie sight.
Then I the Captaine of his warlick traynes
Hoping that he would poyse his daungerous maymes,
with kind [req] rewardę, vnto the court repayrd.
Thinking full little how I should have fay'rde.
where I began, great King your humble vassaile
Maim'de Vis is come to yor Heroick castle,
To seeke releife, I was my leige the Leader
Of yor most loyall hoast, the right succeder
Vnto graund Opis, whome wthout remorse
Dread warrs bereft of life, & did enforce
Him to resigne his place vp to my handę,
That I might be the leader of your bandę.
And that I have it faithfully ꝑformed,
Se this my fatall wound, though wholie armed
My dative case is wanting, which restore
And I shalbe sound, ⟨.⟩as I was before
To wch he aunswers Generall, whose name

483 *prize*] *z* altered from *s* 486 *Reakę*] *e* preceded by mark in ink of hand W 490 Act^{9} 3^{9}. sc: 3.] *sic* (for III.i) 505 *enforce*] *o* altered 506 *vp*] *p* altered from *t* 511 ⟨.⟩*as*] letter ? *h*, erased, tail touches *s* in *as*

Doth tell thy nature, & whose glorious fame
Thy worth doth ꝑallell, thy deepe=dead wound
Doth passe o^{r} weake abilitie to make sound.
And soe deꝑtes. *ffumꝰ*. whie this was freindlie spoken,
A signe of amity, and freindshipes token.
vis. Hyena=like he [feircelie on] weepes at o^{r} distresse
ffor further mischeefe seeking noe redresse.
Fum: More Lyon=like he feircelie on me gazd'e, [FOL. 125b]
Soone dasht my hopes, & made me all amazd'e
ffor when I did begin, (my Leige) poore ffumus
A lack=limme souldier to bould to ꝓsume thus
Tenders vp to [his M] your Maiestie his ditty;
That on his Daungerous woundꝭ you would have pitty.
ffor though through patience ffumus doth endure all,
yet wantꝭ he totally his number plurall.
Affect you plurallꝭ quoth the King? affection
Bring forth effectꝭ of small or no electioñ.
Wherefore Vile Varlett, cease of thy petition
Thinking no more of anie restitution.
Ali: We both for divers dishes had one [sause] sauce,
ffor when I put vp to his grace my cause,
He askd' me who he was, whom vilest peasantꝭ
would dare to interrupt wth their base presentꝭ.
Of fond-complaintꝭ. ffor know although I love yee
Non vacat exiguis rebus adesse Iovi.
Besides quoth he the case you call *vocandi*
Is very often causa exclamandi.
wherefore as in our articles we declard'
ffrom those that have it lost, it shall be barrd'.
Others y^{t} keepe the same we give free will
To vse the same, but wth far greater skill
Soe we dismisse you, aliquis away,
And so dismist I did no longer stay.

536 *fond-complaintꝭ*] hyphen perhaps penmark 539 *often*] *n* altered, ? from *nc*

vis. His good successe makes him to tyrannize
we'ele serve tooth outward, & with Ironi[z]es
we'ele say, all haile to royall princelie Nomen
Although we wish a halter were his Omen.
Ali: w'eele plot some meanes to gett him from his throne
Though with his Dearest life & our deepe moane;
But pray be silent, harke, I heare a humming
Sure by the trampling some bodie is comming.

Act^{9} *3: sce: 2*

Enter volo, fero, dic.

Stand on yor guardę what be you freindę or foes,
yf freindę we thus salute, yf not with blowes
We intercept you. *Dic.* in peace freindes, foes in War,
When as great Verbũ & the Nowne did iarre.
Fum: yf you be freindę, then freindlie give vs notice
Say, neede yee our helpe *Gentlemen*? *volo. yes* so t'is.
We three Dic, volo, ffero, men at armes, [Fol. 126a]
Souldiers to Verbum, sharers in his harmes,
Though not halfe sharers in his good successe
As doth our dolefull misery now professe.
which thing provokd' vs forth from him to flie
To seeke our fortune & ꝓsperititie.
Ali: what saiest thou ffumus, heer's meate for o^{r} Mowing,
Heere's corne ripened for vs of others Sowing.
Brave souldiers we doe much commend your valour,
And would not have you thinke that vnder collour
Of flattring freindship; though you be but strangers
That we prætend your damages or daungers
We were sometimes of note: though now through fate
The onlie obstacles of wisemens state.

548 *we'ele*] ^{2}*e* altered 551 *moane*] *a* interlined over caret 556 *be*] *e* altered 558 *peace*] interlined above caret and before *others* in Hand W 561 volo.] ^{1}*o* altered from *i* 569 *an-* interlined above 572 *but*] *t* altered from *s* 575 *obstacles*] *c* altered from *i*

We are cast downe into misfortunes prison
A Dungeon full of misery & derision.
volo. ffortune's ffooles furtheresse hath put vs to gether,
In her blinde scrowle, as t'were birdę of one feather.
To whom an equall Destiny is allotted,
But yf you follow me that I have plotted,
In spite of fortunes teeth shall make vs knowne.
W'eele goe to all the schoolemasters in towne
And there demaund our long defected ꝑtes,
wch yf they cannot aunswere by theire artę,
w'eele soe turmoyle them, that the Realme shall know,
That we will live in fame. vis. we doe allowe
Thy rare inventions. heer'es one (Aliquis)
Talkę to some purpose, what saiest thou to this?
Ali: why we will leave our former ꝓposition
And onlie give consent to his pætition.
vo: what say yee gallantę? doe you give consent?
Omnes. We doe. vo: then lett'ę be gone, 'fore day be spent./

Exeunt omnes.

Act: 3. sce: 3.
Enter Absurdo ye Pedler of Barbariá.

Thus while I sing along my dittie
Each commendę it to be pritty.
My wares for currant still doe passe,
Absurd in more request nee're was
Come Buy my trinkettę great & small
ffor a little monie take them all.
But now (Absurdo) cease thy pleasant tune
ffor surelie we are iust come to the towne, [FOL. 126b]
Heere must I seeke a man of worthy fame
I feare to many of you know his name
It's Mr Ignorance, his ꝓperties

576 *prison*] *s* begun as *z* 590 *ꝓposition*] *s* altered from *t* 593 *then*] *h* altered from *e*
597 *I*] altered from undeciphered letter

I have ad vnguem, oh hee's very nice *Enter mr Ignorance*
He'es complementall, neat as anie Oyster,
Hush don Absurdo. bugȩ wordȩ, heer's a Royster,
Stand close, & listen. *Igno*: *proh deum atqꝫ hominũ fidem*:
Its a cruell trouble not to teach ad *Idem*.
I thinke in conscience, all the country teachers,
Doe want my Methode, pish. they all are peachers
Ni[c]hil est, si ad me comparatur,
Each one comparde to me is but a prater.
I aske my boyes [w] how they decline Creusa,
The'ile straitway say it is declind' like musa,
Then I goe further, posing them in *Tibur*
The'ile forthwith aunswere t'is declinde like *liber*.
Had they but little more capacitie
Ild'e quicklie send them to the Vniassitie.
Abs: This is the man, his talke doth him bewray,
well I'le to him, t'is hee that is my pray.
Good Mr Ignorance, for soe I think's your name
Please you to buy verbes tences on[a]⟨c⟩eof fame
Cumbo and *Tulo*, and fine new coynd' wordȩ,
wch you shall have as cheape, as prise affordȩ.
Ign: whence com'st thou fellow from Italia?
Abs: Noe sir, not I nor yet from Gallia,
But from Barbaria. *Ig*: ô sett downe thy wares,
Then thou hast noe Italionate affaires
wth which my kindred ever more were troubled
Tell me the troth thine owne prise [m] shall be doubled.
I doe not love this strange Italionating
It is afancy past my imitating
Heere Honest fellow, pray give me thy trinkettȩ,
Heers coyne sufficient for to buy thy iunkettȩ

617 *Creusa*] *usa* perhaps italic 618 *musa*] perhaps italic 623 *talke*] *l* altered from *k*; ? *k* altered from ? *a* 626 *on[a]⟨c⟩e*] *a* cancelled, *c* added but blotted 627 Cumbo] *u* altered from ? *a*

Abs: Thanke you good Master; ffoole & that in graine,
when they are scantie, you may goe for twaine.

Ig: Farwell good fellow. like a rare Physition
I'le of these simples make a composition
Now possum, volo, & th i'rregular traine [FOL. 127a]
I'le bring to fashion, vsing little payne
The out=lawes questions I will soone resolve
Though *Sphinx* (there's reading) doe the same involve.
But I forgett, my Boyes have me expected,
I'le straite to them, they must not be neglected.

Exit.

Chorus.

*Hi*st, peace my Masters, this will grow to treason,
& yf it be not look't vnto in season.
Did you not see,? the traytors had agreed
wth strife to make o^{r} Grammer kindgome bleed.
They sweare, they thretten, scarcelie had they spoken
When enters Mr Ignorance, a token
Of their disturbance, whom wth their defect
They have sore troubled; we must not neglect
The puñishm^{t} of these out=ragious feindȩ.
Sitt then a while, you that be good Rules freindȩ.
And then Parenthesis thy selfe convert,
ffor I must looke vnto my other ꝑte.

Act:4 Sce: j. *Enter Oratio, verbum, Nomen Participiũ, Lillie, Priscian.*

Ora: Now I begin to feele my former riches
when concordȩ be observd' in all o^{r} speeches,
when that the Nowne & verbe admitt noe iarr;
In Number or in ꝑson, oh this War
Hath made our kingdome suffer manie crosses,
& to my sorrow each hath borne some Losses.

653 *see, ?*] *sic* 654 *kindgome*] *sic* 665 *riches*] dot for *i* over *r*

ffor when each private subiect was soe bolde
To catch into his fingers what he could
I know some Nownes, wch now I will not mention,
who have snatcht' vp a dou[b]ble folde declension;
I could name Domus, but I'le say noe more.
As they be thus enritcht some be made poore.
Aptotę I know there be above an hundred,
wch from the fight indeclinable blundred,
Besides some Diptotes, wch the wars did vndoe,
That now *vix quatuor casus tenuere secundo*.
Some Triptotes they return'd wth ⟨.⟩ onlie three
I thinke these losses were enough for thee.
Besides some pronounes wch doe make a shift, [FOL. 127b]
To live in creddittt, yet they wantę their fift.
Noe lesse is yor losse *verbum*, out alack
How I do pittie Dic, *Duc*, *fer*, and *fac*.
with Volo, furo, & that maymed crue
Would not this greive one *Verbũ*? how say you?
Long have you fought, & fearce, but gayned nothing,
Only some Thumpes & Blowes, is [it] it not soe King?
verb: we fought (dread Queene) to long I doe confesse,
I had some daunger, *Nomen* had noe lesse.
All wch redoundę to the, thou had'st the smart,
But talke no more of that, lett'ę take vp hart.
Remember't [that] not, though bloud our battaile cost,
Lett'ę say whatę won is won, what's lost is lost.
yf we recount, & number them soe much,
It will but greive vs, y^{t} our harm⟨s⟩ are such.
Ora: Oh stay, we must consider them a little,
we should not Doe amisse to build a spittle,
ffor our maym'de souldiers somewhere in the citty;
To see soe manie daylie halt t'is pittie.

674 *dou[b]ble*] *o* interlined above caret 690 *is*] interlined above caret 694 *talke*] *l* altered from *k* *vp*] *v* altered 698 *harm⟨s⟩*] *s* altered, ? from *w* or *a*

Nom: It is indeed, & my men they complaine,
Wishing that each then had binne outright slaine.
Rather then all their life long to lye bedrid
ver: T'were fit indeed, there case should be consid'red,
But they ꝑsume soe much vpon their merritt
As yf that from our handę they strayt would teare it.
Craving Rewardę forsooth for all their woundę,
Thus still our pallace wth their clamours soundę.
Fer would have fere, [Bt] but I'le *fferas* make
to serve his turne, some wrong he needę must take.
And Dapper Dic his Dice proudlie claimes,
In gen'all for rewardę tendę all their aimes.
Which Ile cut of, I'le make them for to feele,
That *Verbum* can be angry yf he will.
Nom: you're delt wth by yor men, as I by mine,
One comes, & cries my dative case resigne,
An other saies, his plurall number's wanting,
An other yoapes, his Vocative lies panting,
Amongst King Verbum's campes all drencht' in bloud.
Ha, ha, ha, ha, oh it would doe you good.
To heere what musick all their yawlingę makes. [FOL. 128a]
Some Nownes their Genders, some their numbers lackę.
But all shall fare alike, I doe not care.
I'le give noe more then what I well can spare.
Lillie. Oh be not Tyrantę, good my Lordę be quiett,
Be[e]leeve me you shall gett no honour by it.
you should change [coullers] colours like Camelions,
ffancy all fancies, fearing lest rebellions
Should chaunce for to invade yor peacefull kingdome.
Oh, this is not the readie way to bring downe,
your giddie subiectę vnadvisd' outrages:
Be rulde' by me, requite them wth stout wages,

712 *serve*] *v* altered from *w* 717 ^{1}*by*] interlined above caret

And first speake peace, seeke soe to quell their charms,
yf they denie it, whie; then take vp arms.
Prisc: Noe more of that good King, take not that counsell,
yf war's contynewe, I shall be knockt' downe still,
I could endure a palt but now and then
fflung at my coxcombe by [a] vnskillfull men.
But when each fellow bangę me wth his wasters,
Some breake my head, at lest put by my playsters
How can I live, I prethee doest thou know?
Lill: As yf that thou't dy with a little blowe,
Besides Anomalaes & Hetroclittę
May fight yet never hurt thy head, or wittę
Nay speak false latine, thou committest noe sinne,
And yf thou doest it but ꝓ *Antiptosin*.
Thou by that figure maist put case for case
Then doe you ill, my iudgm^{t} thus to crosse.
Part: Good Sirs be freindę, from wordę growes blows at length,
Pray lett vs loose noe more of Grammers strength.
I know that these late wars enough have tri'de you.
Then live in freindship, & in peace as I doe,
But whie talke I, I know yor disposition;
Thou (*Lillie*) can'st not disagree with Priscian,
who[m] loves oratio, loves him, as his life,
ffor they be in a manner man & wife.
Let peace & concord be amonge vs. *Lillie*. stay there
Good Participle, I am foes wth neyther.
I nee're was out with him, nor yet fell from her [FOL. 128b]
As well appeareth in my latine Grammer.
Ora: I know thou didst not, yea thou wer'st a meane,
To bring to concord all our troubled Realme.

741 *wth*] *t* altered from *c* 746 *hurt*] *r* interlined above caret 751 *growes*] *w* interlined above caret 757 *who[m]*] beginning of *m* deleted

Act^{9} *4. sc: 2.* A noyse within crying.

vox Persue the raskalles so hoe, follow, follow;
intrat Knock downe the Traytors, *Fero*, *Dic*, & *volo*.
Ora: How now, what meanes this vprore in our Court,
vox. See, I am vp vnto the knees in Durte,
To come to tell yor Maiestie the newes,
Out=lawes there be, that doe yor Realme abuse.
I have scant breath to speake the thing at large,
wch was eare while committed to my charge.
Par: What out=lawes quoth hee, heere will be some knockę,
Say what'ę thy name?. *vox*. Oh sir my name is *vox*.
Par: what is thine office, come a little nigher,
vox Mine office Sir? I am her Graces cryer.
Ora: Oh I remember him, *vox* doe not feare,
But lett thy sound come to our princelie eare.
Say whatę the matter? out=lawes thou did'st mention,
what doe these out=lawes? do they breed Dissention?
vox. Alasse Dread Soūaigne, our country swarmes
with verbes Defectives, wch be vp in armes
And caselesse Hetroclitę, wch every day
Abuse yor subiectę, in the kingę high=way
Some Aptotę, Triptotę, monaptotę, & others,
Are wth those verbes anomalaes sworne brothers,
And have conspir'd to gett what ear' they need,
Or they will make your country subiectę bleede.
They chaunc't to fall vpon one Ignorance
whom they have putt into soe strange a traunce
That he is allmost madd, one keepes a stirr
ffor his imꝑative, then enters *Fer*
And sweres he will have *Fere*, & ꝓtestę
Hee'le fferrett all your subiectę from their nestę,
If they will not restore it; volo cries

767 *intrat*] below the line 772 *thing*] *i* altered from *e* 775 *name?.*] *sic*
776 Par:] above the line 796 *If*] *f* altered from *y*

where's my imperative? yf you be wise
ffind it you're best; soe is my worshippes pleasure,
Poore Ignorance he scarce hath anie leasure
To teach his pupilles. More there be some Nownes, [Fol. 129a]
which doe each Day & houre this ffellow trounce.
I hope for these same traytors reformation,
yo'ule call them all to theire examination.
Ora: Doubtlesse we will, for sure we must not sub[b]orne,
Such raskall villaynes, wch be growne soe stubborne.
Nomen & Verbum, we do thinke it fitt,
That you doe send some Herrauld wth a writt.
To fetch these rebelles[,] in, ill should we fare
If of our subiectꝭ we should have noe care.
ver: Madame. we will. *Nom*: soe standꝭ it with our minde,
Ora: Lett then the ꝑson straytway be assign'de.
ver: Speake thou grave *Lillie*. noune & verbe beseeches,
which is the fittest of all ꝑtes of Speeches
To be our Messenger to this rude people.
Lill: My iudgm^{t} maketh choyce of Participle,
My reason is, that he seemes somewhat lusty,
Besidꝭ I take him to be very trustie,
And for because he takes ꝑte of you both
I thinke his Lordship would be very loath
ffor to seeme ꝑtiall in this enterprise,
Eyther for Nowne or verbe in anie wise.
ver: Thy choyce is good, but say, how lik'st thou of it
Don Participle, t'will be for thy ꝓfitt.
Par: I sweare by all my ꝑsons, genders, cases,
The Participle willinglie embraces
The Office, wch you have committed to him
And will ꝑforme it, yf it should vndo him
This arme, wch for your goodꝭ is allwaies readie,

798 *you're*] apostrophe possibly part of tail of *y* in *my* (l. 797) 815 Lill:] above the line *Participle*] *r* altered

Shall take those traytors, wch are growne so headie.
I'le be as true in this same deedę ꝑforming
As steele it selfe, I care not for their storming.

Nom: why that was spoake like Hector that stout Phrygiã,
But heare me, thoushallt' carry with thee Priscian,
You may consult with him, he may doe good,
And bring the traytors in, & shed no Bloud.

Pris: Spare me my Lordę, for I can give no counsell
ffirst will I runne my head ageinst the groundsell.
Ear' I'le to wars ageine, you know with parting, [Fol. 129b]
I burst my head soe, 'thath not yet left smarting

Par: Then I & vox alone will travaile, come on,
w'eele to your pallace all these traytors summõn.

Exeunt Part: et vox.

Ora: Thus shall we bring our Realme to peace at last
when all these same Defectives be laid fast.

Act 4 Sce: 3. *Robinson knockes at dore.*

Priscian who knockę, see that none enters in,
what ear'e he be, vnlesse he be our kin.

Pris: I know not what he is, what shall be done[?] *Goes to the dore*
He saies his name is Robin Robinson. *to aske whose' there*
then returnes.

Ora: Lett him draw neere, oh you are welcome Sir,
What have you [late] harde of this late wicked stir?

Robin: what meane you 'mong the maymed verbes & nownes?

Ora: The very same; they trouble all our townes
And villages, about our grammer=Land,
No place is free, whereof we have commaund.

Rob: I heard of it, & you are she I sought for,
You are the very *Queene* that I tooke thought for.
ffor in good sooth when first I har'de this rumor[e],
I was affeard that this would trouble you more,
Then the late Discord, wherfore being [readie] ever

833 *thoushallt'*] a line separates *thou* and *shallt'* 841 *traytors*] ^{1}*r* interlined above caret
848 Goes] *G* altered 852 *wicked*] *e* altered, ? from *c* 859 *rumor[e]*] ? *e* deleted

Readie to healpe you wth my best endeavour,
I thought for to present to you my service,
But now I see my hay comes after Harvesse'.
Heer's healpe enough your number is compleate
Nom: Nay good sir stay, for you shall take your seate,
Ora: Sitt heere, pray be not coy, for will yee, nill ye,
your place shall be next vnto Mr Lillie.
Rob: We thanke your highnesse; *Ora*: Come leave of these thankȩ
Let vs now studie, to suppresse these prankes.
ffirst when our Messenger hath brought them in,
Lillie shall wth thanomalaes begin,
Them shall he bring into some rule & order,
wch being Done, we'ele cause that our Recorder
Shall come, & read the Hetroclites citation,
Of all wch you shall have the Domination.
what say you kingȩ wch governe this Dominion?
Nom: I think it fitt. *verb*: and soe is mine opinion.
Prisc: Noe lesse Doe I. *Ora*: why then [our Counsel] lead on my freindȩ, [FOL. 130a]
Vnto our Pallace, thus our Counsell endes.

Chorus.

ffy heer'es a doe to end this scurvie quarrell,
I scarce have time enough to change my 'parrell.
Betweene each ꝑte, I hope this wilbe mended,
you see the Kinges have promised they will end it
ffor while these out=lawes makes a shew of Braverie,
vox certifies the counsell of their knavery.
Both kingȩ have chose theire man to bring in Rule
Those out=lawes & to make them fitt for schoole,
Lilly & Robinson they whet their wittȩ
To curbe thanomalaes & Hetroclittȩ.

863 *service*] *c* altered 867 Ora:] above the line 873 *into*] *o* not completed 877 *this*] *is* blotted 879 Prisc:] *c* almost closed, resembling *o* *lead on*] interlined above deletion 886 *Braverie*] *v* altered

Onlie remaines the sentence for their doomes,
Till wch be Done be ple'asd to keepe your roomes.
But soft ꝑenthesis thou art expected,
The Participle must not be neglected.
Beside the Epiloug doth now draw on
wch is my charge, & must be thought vpon./

Act: 5 Sce: 1. *Enter volo, fero, dic vis Fumꝰ, and aliquis wth Mr Ignorance.*

Fum: Sure S^{r}, your boyes had need of more direction,
They cannot aunswere vs for our Defection.
Ign: what say you? can they not? *oh stirpem invĭsam*,
ffor this their negligence see how I'le trice 'm.
I will vntrusse them, *sursum* and *deorsum*
ffrom high to low they all shall goe to horsum.
volo: Nay be not S^{r} so moved, as to breech them
ffor they will learne heereafter, as you teach them.
Ign: *yea. yea.* I would you had binne all a mile hence
when as you put my pupilles thus to silence;
But Sirs, why doe you in such doubtꝭ involve yee?
And vrge poore simple schollers to resolve yee,
Si essent inscij, had they insight in it
They would have aunsw[e]rd' quick, you should have seene it
Oh such a one was I, I was so readie
In Latine speech, in Latine onlie said I
I'le tell you what I am so good a Græcian,
I need not come to learne of Doting Priscian.
Dic. Come come, we come not for to heare your preaching
Tell what we came for, or leave of your teaching. [Fol. 130b]
Ali: Good S^{r} in ffreindlie sort say what ye know,
Tell dic the cause why he is curted soe.
Dic. Come dic will make you tell him, vnderstand yee
Modo imꝑativo I commaund yee.

894 *ꝑenthesis*] ¹*s* altered from final *s* form 900 *Sure*] *S* altered from *ff* 910 *yee?*] *?* almost invisible 913 *seene it*] turned up, with open parenthesis, into l. 912 919 *teaching*] *t* altered, ? from *l*

Ign: *Paucis contenti*. peace & be contented,
I'le tell you strait se how I am tormented
with these hard questions, fy on't, I had lever,
Be dulie ffollowed wth a quartan fever.
Fero. Sir Ignorance I see your braynes do travell
About their losses, for the wch they cavill,
But as for me ye need not vex you further,
My Perfect=tence was gone, I ha' got an other.
Ig: Then shortlie ye will grow to better Rule. ô.
Fer: yes. for I had the reliques of Dead Tulo.
Ig: what, what? is Tulo dead? o Durũ *casũ*,
Doleo ex animo out of minde to rase him,
But wherefore *am I* in the schooles a medler?
Sure, sure I was Deceaved by that Pedler.
But I will right my selfe, or it shall misse hard,
Non ego diserto Il'e not be made a Dissard;
Fum: Come, Come though ffero bears his Borrow'd tence
Il'e tell thee sirra I will not goe hence,
vntill thou finde me out my plurall number;
what fellow? Do'st thou thinke that I Do slumber?
Ig: [N] *Monstrũ horrendũ*. what your plurall lacke yee.
Me thinkę you should not, more do take Tobacce'.
Fu: Why so, I know't, & find them, or I'le smoake yee.
Ig: *Bona verba quæso*. would your smoake did choke ye.
T'will choake vs first, woo't they might never stir more
That thus. torment me. *Aliq*: harke how he does murmur
we'ele vex him worse, you know sir what the Law is
why should I want my vocative? *Igno*: *quid ais*?
Why should I want my vocative? *Ign*: *væ misero*
I would heere were sage Solon, or sweet Cicero

935 *to*] *t* altered   940 ²*Come*] *C* altered from *s*   948 *stir*] extra stroke on *i* creates a minim   950 *sir*] *i* blotted or altered   *what*] deletion above *h*; word may be altered from *w*^{t} 952 *Why should I want my vocative?*] speech prefix for *Aliquis* omitted   *Why*] *W* altered from *T*

Or maior Cato, or some Maior in scarlett,
To pacifie this rude vnrulie Varlett.
Ali: What heare ye not? faith weele shake vp you spirittę
Must we not be regarded for our merrittę?
We were all souldiers in these bloudie broyles,
Some hurt, & maymed, others getting ffoyles
Thus loosing limbes, & having cranies crac't,
We come for remedie for our defect. [FOL. 131a]
Ig: ffaith Sirs I cannot, but condole yor chance,
To seeke for remedie of Ignorance.
Ali: O Sir your pleasant wordę shall never tame vs,
we'ele not be aunswe'rd wth an ignoramus.
vis. Let me alittle question him for my case,
where is my Dative. *quæso magister dicas.*
Ig: why; in the wars you were of him bereaven;
yet *forte sir*, it may be you may have him.
vis. May me noe Maies, nor forte me no forte
As I am vis I will h'ate. Ile be short wee' ye.
you will not speake, you long to have a clap?
Take that then soe, take vp your wisedoms cap.
Ali: Nay, though ye mock him Do him yet no wrong
Stay Mounsier, stoope not, I will put him on. *putę [on] the fooles cap.*
Now are you fitted. *Ign: o ridiculi* *on[e] ignora: head*
They vex me mock me, whether shall I fly?
Exit ignorance.

Act: 5 sce: 2.

Enter Particip: [w] and vox with a watch.

vox. Now sure, We mett the very man, we spoke of,
Those out=lawes ar not far, their sport's new broke of.
Par: Loe heare they are, lay hold on vis, & Volo,
Two sturdie varlettę, see they scape not, so loe,

966 *alittle*] a line in hand H separates *a* and *little* 976 Ign:] *I* altered from *O*
978 *They*] *e* partly missing

Now for the other. *vox. fumus* he is vanisht;
I would all such as he were cleerelie bannisht',
Now take the rest, & bind them, *Fer*: *Fero yeeldes.*

Par: Heere is a coyle indeed, all grammer feildes
Doe swarme wth Hetroclites, all places cry out,
vpon their misbehaviour & rioutt
This shall be mended, you must all appeare
At your kingꝭ Pallace to be iudged there. *Exeunt omnes*

Act: 5 sce: 3.

Enter Priscian wth two supplications.

O misery, thus still to be abused,
My head [sh] still knockt' my plaisters daylie loosed;
ffresh enemies still, fresh woundꝭ, fresh buffettꝭ given,
Alack I would I never heare had liven;
Or dealt with Nowne or verbe, or grammer=stuffe
They strike, as yf my head were hammer=proofe.
It greeves me, others should be well, & I sick
But that I hope I shall have better Physick [FOL. 131b]
Be comforted & cheere thy selfe (old Priscian);
Though thou bee'st sick, thou nee'r shall want Physitian,
Oratio is my frend, & manie moe,
Nay I may say I have not anie foe;
None that will hurt me, *nec foris nec domi*,
But newcome fellowes, those that do not know me.
ffor these & such enormities ageine me
I will vnto the Grammer kingꝭ complaine me
And for that purpose heere I keepe my station,
To offer to them each a supplication.
I heare expect my time, but ear', or late
Suters must Daunce attendance at the gate.

990 *cry out*] a faint line is drawn between these words 991 *rioutt*] *tt* blotted, possibly obscuring punctuation 993 Exeunt omnes] between ll. 993 and 994 996 *thus*] *t* altered from start of *s* 1015 *Daunce*] *D* altered from *a*

Act: 5 sce: 4.

Enter verbũ, Orão, Nomen Robins: & Lillie.

ver: Now learned Lilly, since those stearne anomalaes,
Have raysed fearefull tumultȩ, & gone from my lawes;
Breaking the civill peace throughout our kingdome,
Let this be thy charge into Rule to bring 'm.
Make statutȩ, such as may restrayne them dulie;
Ill fares the Realme, where subiectȩ be vnrulie.

Nom: ffor Hetroclitȩ to Robinson I graunt,
The like commission for theire restraint.

Lilly: Both shall be done, but see heer's Priscan,
It seemes, he comes wth some petitian.

Ora: welcome good freind, how does thy scalp of late?
Thou never go'est wthout a playstered pate.

Pris: Thus it fall's out, some enemies waylay me,
Such enemies, as I know not nor they me.

No: Hold the thy papers Priscian, our meeting
Hath binne thought fitt of for a generall sitting,
In iudgm^{t} of some vagaboundȩ; thy writing
will serve for evidence in their inditing.

ver: But my Anomalaes, are become so rough,
The'ile plead not guilty, though we have good proofe.
Say, they be stubborne, how shall we constraine them;

Lillie. Some verbes deponentȩ heare do lay ageine them̓,
I have their depositions reddie written,
wch shall be read anon, when you have sitten.

Rob: And heere ar articles for the Hetroclites.

No: T'is well, we'eele tame those hawtie Neophytes
And bring them vnto Rule & and due submission
Having both article, & deposition.

1016 *4*] altered from *3* 1018 *learned*] interlined over caret 1019 *fearefull*] beginning of another *f* after ^{1}*f* 1020 *Breaking*] *r* altered, ? from *e* 1026 *Priscan*] *sic* 1033 *binne*] missing one minim 1037 *proofe*] ^{2}*o* partially obscured by *f* 1039 Lillie.] above the line

But see, hee'rs Participle or I mistake him, [FOL. 132a]
With that vnrulie crue that hee hath taken.

ver: Then lett vs sitt in iudgm^{t} as beseeme vs,
Come faire Oratio, take thy place betweene vs.

Enter Participle vox, wth the anomal: and hetro: after all be sett Rob: as clarke (wth Lillie) thus beginnes.

Rob: *Cryer* make an oyes. 3. All manner of ꝑsons singuler or plurall
wch ow suit, or have binne bound over to the Court: of the 2
Mightie potentates Nomen & verbũ make your ꝑsonall apꝑance
eu͛y one as hee's called. vpon payne & ꝑill that will fall
there on.

Lillie you[r] ar not ignorant (vnited freindeꝭ
To grammers civill orders) to what endeꝭ
These mighty princes cause our meeting heere
Dayning them selves in ꝑson to appeare;
ffor why? heere ar presented to our sighteꝭ
Lawles anomalaes & hetroliteꝭ,
Heare to be censurd for their outragis,
In which their censure w'eele not be remisse
nor on the other side, will we be cruell,
only defend o^{r} righteꝭ, so we hope you will.

Rob: we will Doe all in Rule, nothing in fury,
Nor will we stand t'empannell any iury.
Cryer say after me. All you hetrocliteꝭ, suꝑant or deficient
Aptotes, Monaptoteꝭ, Diptoteꝭ, & triptoteꝭ stand forth, & heare y^{e}
tarticles layd ageinst you.

No: How comes it so few hetrocliteꝭ appeare?
Say Participiũ [th] *Par:* they ar fled for feare.
But yet I hope they shall be tane at leasure,

No: yea, & we'ele make them all attend our pleasure,

1052 beginnes] missing one minim 1054 *ow*] *sic* *the*] *e* altered from *is* 1055 *potentates*] *a* altered from *e* *make*] *k* altered from *y* 1072 *tarticles*] *ta* altered from *a* 1074 *they*] *e* altered from *a*

But where is ffumus? Sir he 'scaped me,
He'ele still be out of Rule I plainelie see.
Ro: To ꝓceed W^th these Hetroclitꝭ heere ar 3 articles framed ageinst them.

* * * * * * * * * * * * *

what say you guilty or not guiltie.
vis. guiltie, but pray consider you my case.
No: Nay such as you merritt but little grace.
Al: I guiltles am, I hear crave but my due,
ffor aliquis behaves him selfe most true.
And for the out=lawes I was not amid them
yf anie tumultꝭ were t'was nemo did them.
No: O grosse vntruthes! how is this aunswer'd here?
vprores ar raysd, yet aliquis not there? [FOL. 132b]
Ora: And so of nemo he complaines vntrulie,
ffor of all Hetroclitꝭ hee's the most rulie,
That he begat these stirs Il'e not beleeve,
Hee's not in case, he wantꝭ his genetive.
Nay we may rather make him gram̃ers Dearling,
in favoring him, as he doth favour learning.
The good we had of late he did ꝓcure all,
He is our onlie freind, admittꝭ no plurall.
what ṣhould I speake more of his com̃endation,
He onlie 'mongst all others [ke] lives in fashion,
keepes the ould fashion, newcomes still detesting,
And labours still in goodnes, never resting.
ffor wch I doe conclude, that nemo's hee,
that *fælix ante obitam* shall bee.
ver: Come Priscian we doe now accept thy writing,
Thou art misvsd, we'ele labour for thy righting.
Read these his supplications Lilly.
Lillie. * * * * * * * * * * * * *.

1077 *ffumus? Sir*] speech prefix for *Participle* omitted 1096 *doth*] *d* altered from *v*
1102 *goodnes*] *oo* altered

ver: what thou requestest heare is nought but reason,
Those men w'eele bannish, & their wares w'eele sease on.
Lill: * * * * * * * * * * * * * * *.
No: Thankes gentle Priscian & in quiet rest thee,
w'eel punnish those offenders that molest thee
Lill: *Cryer* call the rest. Anomalaes come forth, & heare yor iudgm^{t},
Heere are inditementꝭ, wth some depositians,
All wch in tenour doe agree wth Priscians. guilty or not
ver: The matters all ar [pl] palpable[s] & cleare,) guilty?
As by the former writtingꝭ doe appeare;
Now Lillie doe thou execution on them,
Of this my iudgm^{t}, wch I passe vpon them;
ffirst be they shut, lest that they goe at Randũ,
Within the grammer prison calld' *notandũ*,
Then let them there remaine for ever maymed,
Nor gett what ear' they [we] lost, or bouldlie claymed.
They have deserved worse, we ar not strict;
Lillie. Ile' Doe according to yor grave edict
Nom: So for those Hetroclitꝭ my minde is this,
That they be prisoned for their amisse,
ffirst in regard so proudlie they disdeine vs
Commit them to the Dungeon cald *Quæ gemas*.
There let them lodge, hang chaines, & shackles
on them, [FOL. 133a]
vntill that thingꝭ with six feet creep vpon them,
I meane Hexameters shall be there guard;
vis. *Parce* good King let our complaint be heard.
No: Our iudgm^{te}' past, & may not be wthstood,
you will be naught, though we be neere so good;
Therefore I say in vaine ye cry out parce,
Away wth them. our Law affordꝭ no mercie.

1115 *inditementꝭ*] ^{1}*t* altered from *g* 1116 *guilty or not*] below the line 1117 *palpable[s]*] ^{1}*p* altered from *a* *guilty?*] between ll. 1116 and 1117 1122 *grammer*] ^{1}*r* altered from *a* 1124 *they*] interlined above deletion with caret 1130 *Commit*] *C* altered from *T* 1131 *shackles*] *c* begun as *k*

Ora: Thus is the grammer goverm^{t} in peace,
Thus daungers on grave Priscians head will cease,
Thus Nowne & verbe in quiett rest remayning
Perfectȩ oratio, both together [raygning] reigning.
Thus as we hope, & as we ever ment,
Procure we peace to vs, to all content./

Exeunt omnes.

Epilogus.

The sentence being past, nothing remaines
But execution; stay. heere endȩ our paynes.
ffor our auctoritie it is so slender,
That iudge we may, but punnish noe offender.
Then we would have yee know that this o^{r} story,
Is true, but coverd' with an Allegorie.
ffor when we tooke in hand this toy, we ment
By the Defectives freshmen to ꝑsent,
wch daylie like irregulers rebell
Ageinst vs seniors, pray Sirs marke me well.
Absurd & inorance epitomize
Their huge great volume of ill ꝓperties.
wch be soe true a pattern of that stem̄,
that they belong to none but fooles & them.
The nowne & verbe, Oratio, wth the rest;
I doubt not you enterprett for the best.
Priscian may reꝑsent our misery,
wch Daylie are abused as you see.
But to conclude we ꝓmised to please,
your eies wth warres wch we in quiett peace,
have now ꝑformed, yf it have offended,
Lend vs your handȩ, & it shall soone be mended.

Finis.

1143 *ever ment*] *r* and *m* separated by a line 1155 *rebell*] ^{2}*e* altered from *b* 1157 *inorance*] *sic* 1169 Finis] extra minim on ^{1}*i*

Item 18 in J.a.1: Fols. 186a–200a *Gigantomachia*, or *Worke for Jupiter*

The original title of this play was 'Worke for Jupiter', and it is the annotating hand (Y) which adds the title 'Gigantomachia'. No date is given in the manuscript, but it is possible that the play was written between 1613 and 1619. In *Gigantomachia*, when the revolutionary giants are picking up the hills which are their chief weapons, Thumpapace says (ll. 513–17):

> Come Rounciuall, we're lag, but weel' not hammer,
> Wee'le chuse those hills, that lie besides the Grammer.
> Heteroclits, that are in mappe of Lillie,
> In that his treatise, where he is soe hillie,
> Ile take dindĩmus, and take thou Gargãrus.

David L. Russell notes that the hills Dyndimus and Gargarus are mentioned in the glossary to Lily's *Grammar* (1567, STC 15614.2, sigs. H8b and Ila),[1] and in that sense they might be said to be 'besides the Grammer'. But the hills are not heteroclites, and the introduction of this word, so obviously extraneous to the context, may be a reference of some kind to *Heteroclitanomalonomia*, which is dated 1613 in J.a.1. In the same way, a line in *A Christmas Messe*, dated 1619, may provide the *terminus ad quem* for *Gigantomachia*. At ll. 179–80, King Brawn says:

> This rude rebellion farre more stomack I
> Then did the Gods the Gygantomachy.

The second line here, l. 180, is squeezed in between ll. 179 and 181 in the regular scribal hand. Since it completes the couplet it is impossible to know whether l. 180 is an afterthought, or whether the scribe simply omitted it at first. Even so, it is reasonable to suggest that both references—to heteroclites in *Gigantomachia*, and to 'the Gygantomachy' in *A Christmas Messe*—may have been intended to recall recent performances in the same college, or elsewhere in the university.[2]

Gigantomachia is the fourth Christmas play in J.a.1, the others being *A Christmas Messe*, Jonson's *Christmas his Show*, and *Periander*. In the first Chorus the Knave of Clubs complains to the audience that 'Christmas once past', he is scorned (l. 21). At Oxford, Cambridge, and the Inns of Court, Christmas was the

[1] David L. Russell, p. 139 (edition cited on p. 3, n. 7).

[2] *In Every Man out of his Humour*, v. v. 54, there are said to be 'Goggle-ey'd Grumbledories' who, fed with pork, 'would ha' *Gigantomachiz'd*', or openly rebelled. Jonson's editors describe 'Grumbledories' and '*Gigantomachiz'd*' as words coined for this play (*Ben Jonson*, iii. 581 and ix. 476–7, edition cited on p. 8, n. 4). *EMOHH* was first played in 1599, and first published, in three quartos, in 1600. The references to 'Grumbledories' and '*Gigantomachiz'd*' are retained, without significant alteration, in Jonson's 1616 folio *Works*.

favourite season for card playing, as well as for plays.[1] Although *Gigantomachia* contains no specifically academic references, the Epilogue's description of the piece as a 'mocke=play' suggests that it should be placed in the category of a university 'show' (together with *A Christmas Messe*, which, despite its highly formal structure, might also be taken for 'some mockshow': see above, p. 35).

The play may have been written by a member of the college in which it was produced. The major requirement for a production would be a balcony on which the gods could appear 'aloft' (l. 257). This direction only appears in scene five, but the text shows that the final battle was to take place with the gods ranged above and the giants below. In that final battle the giants appear with hills onto which they climb: 'Pelion lie thou there, whilst I gett vp on thee' (l. 503). In one version, when the battle is finished and the giants are defeated, Bouncebig 'wounded on the top of his hill, speakes' (text of cancel, l. 17), and elsewhere he climbs down from Pelion's shoulder (l. 566). Russell suggests that since the hills are listed in the *dramatis personae*, they were played by actors.[2] Fairly athletic undergraduates could have held each other on their shoulders for short spells, and it would certainly have been easier to manage entrances and exits with men rather than with large stage-property hills. Beyond the balcony and the hills, the play requires some costuming, notably Thunderbolt in a 'furies coate, ougly vizard, / and à wheele of crackers' (text of cancel, ll. 3–4).

The style of *Gigantomachia* is much less polished than that of *Heteroclitanomalonomia*. The language is often rather awkward, and the rhymes are forced: Bellona/alone a, dolphin/all finne, ne're be seene/ George à greene. The author prefers weak line endings, frequently using eleven- or even twelve-syllable lines, and feminine rhymes. Words are invented or adapted for the convenience of the rhyme: 'narrare', 'fagery'. Russell claims that there is a distinction between the rusticity of the giants and the polish of the gods,[3] but Jupiter is still capable of lines such as 'My ribs will burst with anger, fetch à Cooper, / ffor I shall burst except I haue à hoope heere' (ll. 316–17).

Unlike *Heteroclitanomalonomia* or *A Christmas Messe*, *Gigantomachia* does not have a classical five-act structure. It consists of ten scenes with the Chorus appearing as prologue, after scenes 3 and 6, and as epilogue. As usual, the play is based on a debate: here it takes the form of the giants rebelling against the gods, a conflict which might have almost any application. The giants could represent

[1] Hilton Kelliher points out that in 1614 gaming at cards and dice was 'expressly sanctioned in the judges' order for the Inns [of Court]' but that 'the prosecution in the Vice-Chancellor's Court at Cambridge, on 22 January 1609/10, of two fellows of Christ's and Corpus for making sermons against cards suggests that [at Cambridge] too they were regarded as legitimate pastimes during the Christmas period' (private communication to the editors).

[2] Russell, p. 110.

[3] Russell, p. 32.

freshmen rebelling against more senior undergraduates, or a political uprising against an established secular or religious government. There is no obvious political sophistication in the play, although there may be an implied contrast between the giants' rebellion and the one with which the action begins, where Jupiter, Neptune, and Pluto, having overthrown Saturn, carefully share out their spheres of influence.

Several remarks by the Knave of Clubs suggest that the play follows the form of a card game: 'ffor all this while the cards haue beene à shufflinge, / By this time they are dealt, now comes the scufflinge' (ll. 238–9). Although 'Clubs are turnd trumps' (l. 240), the clubs, or giants, do not win. 'Some man will aske me', says the Chorus in scene 6 (ll. 341–5),

> why I should intend
> To let them fall which vse me as theyr frend,
> Gyants and clubs are seldome seene asunder,
> Why should I let them haue the worst you'l wonder.
> O sirs. I am a club, but yet the knaue . . .

As with the plot of the revolution, the card-game structure is not fully elaborated.

The manuscript of *Gigantomachia* measures 150 × 203 mm. There are double ruled margins top, right, and left. On the first three pages the top margin measures 30 mm, and on succeeding pages it measures 20 mm, as do both the right and left margins. There are no running headlines. Catchwords are used except where the next page begins a new scene.

The primary hand is secretary, with italic forms used for headings. The hand is characterized by a certain lack of distinction between *Y* and *y*, by the occasional use of an accented *a*, by irregularly placed apostrophes, and by two different question marks, one regular and one reversed. The same hand revises the manuscript and makes numerous small changes in a different ink, suggesting either that the scribe was the author, or merely that he was accustomed to transcribe at one sitting and correct at another. One peculiarity is the presentation of mid-line speech prefixes, which begin and end with colons, thus ':Plut:' (l. 380).

The cancel pasted onto 199a contains another hand (see Plate 4). In this version the conclusion is different, with all the giants being killed off (see below, p. 120). This change could have come from any one of a number of sources.

In the textual footnotes below, an asterisk * indicates that the addition or alteration was made in a different ink.

Gigantomachia: [FOL. 186a]

or Worke for Iupiter.

Personæ.

Iupiter.
Saturne.
Mars. }
Pallas. } Gods.
Mercury. }
Neptune.
Pluto.

Bounc-bigge.
Rumble.
Rounciual à Gyantesse.
Thunder-thwart }
Huge-high. } Gyants.
Bumb-cracke. }
Thumpapace.

Olympus.
Pelion.
Ossa. }
Malagolum. } Hills.
Plymlymmon. }
Dindîmus.
Gargârus.

Chorus. [FOL. 186b]

The Knaue of Clubs.

What are ye sate soe close? 'tis wel done, wel done;
Yet I could wish, your cardinge still had held on:
Christmas once past, you scorne the Knaue of Clubs,
I am thrust out of dores, those churlish Chubs
The boyes within there, bid me seeke my fate,
Tel me that cardes are almost out of date;
Indeed the first three Knaues are, I confesse,
But I, know, I, the cheifest of the messe,
Doe purpose for to shew yee one more boute,

0.1–1 *Gigantomachia: / or*] in hand Y (see above, p. 13) 3 *Bounc-bigge*]* altered from *Bounc-bige* 5 *à Gyantesse*] added* 19 *done, wel*], added* 20 *on:*] *:* altered from ,* 21 *Knaue*] perhaps *knaue* *Clubs,*] , added* 24 *cardes*] *a* written over an erased letter, possibly *l*; perhaps the scribe began the word as *clubs* *date;*] *;* added* 25 *are,*] , added* *confesse,*] , added* 26 *messe,*] , added* 27 *boute,*] , added*

Or [t⟨ . . ⟩] two, yfayth, ere christmas once goe out:
Yet, nor at Mawe, nor Gleeke, our play is Ruffe,
where yee shall see the gods, and gyants cuffe,
And now, and then, a game or two at Loadam,
when that the gods at Ruffe haue ouerthrowd' vm./
Sit still a while, and see my noble harts,
How valiantly the clubs will play theyr parts.
I will be Chorus, they'le make quicke dispatch,
Whoe ever meets with clubs, now, meets his match.

Sedet.

Gygantomachia./ [Fol. 187a]

Scene Prima

Alarum, Enter with victory Iupiter,
Neptune Pluto, Saturne bound and
crownd, led in by
Mars and Mercurie.

Iup: Thus with a powerfull arme, your force is queld
And haughty courage: you that could not welde
The weightey packe-horse burthen of a crowne,
Why, we will ease ye; good sirs take it downe.
Thou that to lingringe hopes meanst to reiorne vs,
Weel carue now for our selues, we will Saturnus.

They Vncrowne Saturne.

Tel me with what face canst thou stand to all
That's layd agaynst thee, bloudy Canniball.
Thou on thy childrens entrayles that didst gorge,
Gnawinge theyr mangled limmes, I doe not forge
That which I now haue spoake; thou knowst thy tyrrany

28 *two,*] interlined above deletion with accent *yfayth,*] , added* 31 *two*] over another word, presumably the same one as in l. 28 *Loadam*] altered from *Loaden** 32 *Ruffe*] *R* altered from *r** *ouerthrowd'*] apostrophe slight, far above *d*, close to l. 31 *vm.*] *m** altered from *em*, 33 *Sit*] *S* altered from *s** 34 *parts.*] . added* 35 *they'le*] ²*e* weakly inked, or poorly erased *dispatch,*] , added* 36 *match.*] . added* 37 *Sedet*] *S* altered from *s** 45 *courage:*] : added* 47 *will*] *ll* thickly inked 49 *Saturnus*] *us* altered from *e** 52 *thee,*] altered from *the** 53 *gorge*] tail of ¹*g* blotted

Doomd me to slaughter too; was ever any
The like to this once heard of vilany?
Why? 'tis à greiuous sinne for one to kill any,
Much more the plants, that sprunge from thy owne truncke,
There's none would do'ot I thinke, but beeinge druncke.
Thou knowst' my nurses hired Taberers,
To beate theyr sheep-skinnes, as we hire laboureres,
And all to drowne my dinne, when I had cride,'
Least
Least if thou heardst me, they, and I had dide'. [Fol. 187b]
But, I suruiue, and liue to recompence
Thee for these cruelteys; sirs haue him hence.
Satur: Ô whether doe yee drag me in a garter?
What must I, must I, downe to tenebrous Tartar?
Is this the parte of Sonnes? hath your loue kept tune
With nature's lore, fie Ioue, fie Pluto, ah Neptune!
This is your doeing Ioue, an act fellonious,
Thus to depose thy hoary father Cronius;
If thou hast any sonnes, this be thy curse,
May they vse thee, as thou hast me, noe worse.
Iup: Haue done your groylinge, hould your toothlesse rage,
We haue power, (but [wil not] spare to vse't) this heate to 'swage.
Had nature made thee any, but our Sire,
Thou shouldst haue deerely payd for't, doe ye hire?
If that ye had not, let me ne're be seene,
Had ye beene full as good, as George à greene.
Since tis as tis, wee'l grant ye haue à roome
Amongst the plannets; that shall be your doome.
And now y[e]' 'are old, that ye may tend deuotion,
Thirty

57 *vilany?*] ? altered from :* 58 *Why?*] ? altered from ,* 60 *do'ot*] *sic* 61 *knowst'*] *sic* 62 *laboureres*] ²*e* weakly inked, or poorly erased 63 *cride,'*] *sic* 64 *Least*] *L* altered from *l** *dide'*] *sic* 67 *garter?*] ? altered from ,* 68 *Tartar?*] ? added* 76 *spare to*] interlined above deletion with two carets* *vse't*] ? *s* altered from *l* 77 *Sire*] curlicue on *S* resembles an apostrophe 78 *hire?*] ? altered from ,* 83 *y[e]' are*] *e* deleted to adjust metre* *deuotion,*] , altered from .*

Thirty longe yeares we giue you for your motion. [FOL. 188a]
What power's else, vnto our selfe alone,
As proper, weel reserue, away, [begonne] be gone. /

Exit Saturne

Now since the world belongs vnto vs three,
Let's share it equally, I, thou, and hee.
Our selfe wil be heauens kinge, because ye know,
Our power effected Saturnes overthrowe.
You of the Sea, and of all waters lord;
You of earths netherlands. doe ye thus accord?
Amb. Wee doe. Iup: Then in, where we wil feast a largies
That done, each one, goe thether where his charge is./.

Exeunt Oēs.

Scena Secunda.

Enter the gyants, Bounc-bigge, Hugehigh, Rumble, Bumb-crack, Thunderthwart, Thump-apace.

Bounc. You doughty gyants, shall I speake vnto yee,
Aboute a matter I would breake vnto yee?
If you will promise your assistinge hands[s],
Our powers united, whoe is hee withstands?

Bounc-bigge

Bouncbigge your Lurdane, I, though now made humble [FOL. 188b]
By Saturnes brats aboue, acquaynt yee, Rumble,
Huge-high, Thunder-thwart, Bumbcrack, Thump-apace,
With vncouth chances, that are come to passe.
Ioue that vsurper, that did late vnthroane
His father Saturne, that old honest Crone,
With his tow brothers, Neptune, and Pluto grimme,
His champion Mars, and more, that followe him,

86 *be gone.* /] altered from *begonne** 91 *overthrowe*] [2]*r* added with caret* 98 *Bounc-bigge*,] altered from *Bounc-big*,* 102 *hands[s]*,] comma added after erasure of [2]*s** 103 catch-word *Bounc-bigge*] altered from *Bounc-bige** 104 *Bouncbigge*] perhaps *Bounc-bigge* 107 *vncouth*] *u* inserted above a caret* *passe.*] . altered from ,* 110 *tow*] *sic* *grimme*] horizontal stroke above [2]*m* cancelled*

As 'Pollo, Marcur'y, and his mankinde Pallas,
Doe vse vs earths sonnes, ill; playnly shall as
Apeare vnto yee all, by my narrare,
Breefly without ambages or fagarey.

Huge Out with't braue Bounce-bigg, wee are all a gog,
Till wee may heare, how to throw of this clog.

Bounceb: Knowe then, those haughty gods, are too too boysterous,
Whoe in this little plott of earth would cloyster vs,
You see the earth, why, 'tis noe habitation,
To bound our valours, wee must haue a nation,
Larger then this poore world is, to contayne vs,
The which if you'l consent to, we may gayne thus.
This

Huge This penning of vs vp, soe stomach I, [FOL. 189a]
They'l force vs to a gygantomachy.

Bounc You'r right. I meane to bandy w^{t}h the gods,
Though, you may thinke perhaps, there is greate ods
Twixt vs and them, yet since they beare vs grudge,
There's noe one heere but scornes to be theyr drudge.
Why, whoe is hee amongst vs, not as good,
As any of them all? I'le spend my bloud,
Before i'le beare this rogish slauery,
Goe too, you doe not spie theyr knauery.
Though they be gods, wee be not silly men,
How say yee lads to this, what will yee then,
Consent togeather to effect this plott?
Dare ye adventure sirs? Rumb: pish dare we not?
Doe yee misdoubt our resolution?
I hope we are not of that constitution.
Sit fast proud Ioue you'r best, and Iuno Ioane,

112 *Marcur'y*] altered from *Mercry** 115 *ambages*] detached descender on part of *g* resembles an apostrophe 116 *Bounce-bigg*] altered from *Bounce-bige** 127 *greate*] ^{1}*e* resembles *c* 131 *all?*] *?* altered from ,* 132 *i'le*] *l* heavily inked *slauery*] *r* altered from *y* 133 *too,*] ^{2}*o* added* 137 *Rumb:*] interlinear addition, with caret* *not?*] *?* added* 138 *resolution?*] *?* altered from ,*

ffor Rounciuall and I, bespeake your throne.
Rounciuall my wife, and I her husband Rumble,
Doe meane (take't as you wil) both downe to tumble.
Thump: By right

Thump: By right of conquest, wee may all depose, [Fol. 189b]
Soe Mercury, Ioue's heire, wee'l wipe your nose.
Tut, neuer tel vs of your Caduceus,
Light nimble=heeles, wee'l fetch our Briareus,
w^{t}h's' hundred hands, to clog your hands, and feete,
Though Iupiter looke on, thus I thinke meete.

Thund. This hand shall plucke vp Pelion, and Ossa,
And throw them to Olympus, as wee tosse à
Light tennis ball, weel' front them in the spheares,
And each man pluck à god downe by the eares.

Bumb: Though I speake little I'le not be a Cipher,
'Tis not à petty gods stroake, that I'le flie for,
Wee will oppose our selfe in ful careere,
Agaynst the strongest, and I care not where;
And though I say't my selfe, I'le cope aloane,
With Ioue himselfe, wett in the burninge Zone.

Huge. Bid them defiance. Bounc. Gods beware your Cruppers,
Weel haue y^{t}[our] greene cheese tosted for our suppers.

Huge. Tis not your proud lookes, that canne turne and winde vs,
But yee shall knowe, yee come an are behinde vs.
Exeunt ōes

Scena Tertia. [Fol. 190a]

Enter Mars, Pallas, Mercury
Iupiter

Iup: Ye champions Mars, and Pallas of our kingdome,
I am t[o]''acquaynt yee lately with à thinge done,

145 *Mercury*] *M* altered from *m** 147 *Light*] *L* altered from *l** 148 *w^{t}h's'*] *sic*
150 *Ossa*] *O* altered from *o** 152 *weel'*] *sic* 159 *wett*] ^{1}*t* altered from *e*
160 *Cruppers*] *up* covers *au**, originally *Craupers* 161 *y^{t}[our]*] *t* added*

Since Saturne ⟨l⟩eft y^{e}[e] crowne, which now diuided,
Is by our selfe, and by our brothers guided,
In our three regions, which by portions even,
Are Sea, and hel, for them, and for me heauen,
With the adiacent, and contiguous earth,
A portion more then theyrs, by right of birth.
Neptune his subiects, they, obserue his lore,
And bound themselues in bankes, for runninge o're,
And when his herauld, the blew-mantled *Triton*,
Giues but his summons once, they all runne right on
The wayes of loue and duty, as say authors,
To gratifie [and] or grace theyr kinge of waters.
My brother Pluto hath à greater charge,
His teritories beeinge soe darke, soe large,
Yet for ought I heare, Cerberus and Charon
Order his goblins soe, that neuer dare one,

Amongst.

Amongst them all at's gouerment repine, [FOL. 190b]
Or studie to depose him, as doe mine.

Pallas & Mars
startle.

Tis too true Pallas, and thou noble Mars,
I am opposed by the vniuerse,
That wer't not for your peace, (except my wife)
Ioue would bee quickly weary of his life.

Mars Ambitious varletts doe they seeke our kingdome?
Cannot the earth contayne them? I wil bringe downe
Theyr lofty stomachs, I am past all patience,
If that I come, I'le bringe them that shall chase ye hence.
Vulcan my armour, fury bringe my horse,
Employment's toward, now heere's worke for Mars.

170 *y^{e}*] altered from *ye* *crowne*] perhaps *crawne* *diuided*] ^{1}*i* altered from *e** 181 *or*] interlined above deletion* 185 *goblins*] *o* altered from *a** 188–9 Pallas & Mars / startle] added* 190 *Mars*] *M* altered from *m** 194 *kingdome?*] *?* altered from .* 198 ^{1}*my*] extra minim in *m* used to begin *y* 199 *Employment's*] *E* altered from *I**; apostrophe appears to be part of a loop at the end of *s*

Pall: Pray take me with you, neuer goe alone a,
Ile be noe longer Pallas, but Bellona.
Mer: Heauens Herauld Hermes doth his seruice tender,
To stop this mischeife, least it more should gender.
Iup: Your loues, our thankes: but doe ye knowe the brood
Of these insultinge rebels? it were good
I did acquaynt you. They are sonnes à hundred,
The snake=foote sonnes of earth, I heard ''vm numbred
How mou'd
How mou'd,' how hartned to this fowle attempt, [FOL. 191a]
I know not, I, vnlesse vppon contempt.
W^ch^ if your power can ought, or ought can I,
Were they myne owne sonnes they should deerly buy.
I sweare. Pall: O doe not: Iup: But I will by gis,
Pall: That's not the oath, that I did feare I wisse.
Iup: Persist they, I stand bound wth wrathfull spleene,
And fiery indignation, sharpned keene
W^th^ whetstone of revenge, neuer to slake
My angers fury, till the earth I make
An vnfrequented desert, and the brood
Of earthlings all doe perish wth à flood.
By stix I sweare, that dreadfull hellispont,
And that oath's deepe enough or else the diuells on't.
Amongst them all, there shall not one rascalion
Suruiue, but honest Pirrah and Deucalion.
Pal: They are but rightly seru'd, on iust condition
W^th^ water for to [purge] quench theyr hot ambition.
Iup: Goe Mercury, and both our brothers summon,
In our name will them, brauely for to come on,
W^th^ all theyr kind subsidiary forces,
Munition

204 *thankes*:] *:* altered from ,* 207 *heard ''vm*] *sic* 208 *mou'd,'*] *sic* *to*] interlinear addition, with caret* 211 *deerly*] *r* resembles *e* 213 *wisse.*] . altered from , 225 *quench*] interlined above deletion

Munition, victuals, armour, men, and horses. [FOL. 191b]
Mer: I goe; and haueinge greate Ioue for my warrant,
I shall not spare to doe the rebels arrant. Exit Mer:
Mars I could desire the hecatombe of these,
Were only sacrific'd, thy feirce wrath to 'pease.
Pall: Forbeare to motion't, his oath's inviolable,
To make him breake it, you nor I are able.
Exeunt oēs.

Chorus

ffor all this while the cards haue beene à shufflinge,
By this time they are dealt, now comes the scufflinge.
Clubs are turnd trumps, the gyants haue the ace on't,
Yet shall they loose, for all they sett a face on't.
ffor though the gods haue ne'[e]re' à club in hand,
They'l' winne the game, as yee shall vnderstand.
Yet when Ioue's wrath doth on the gyants light,
I'm sure, you'l deale wel w^{t}h the clubs this night/.
Sedet.

Scena Quarta. [FOL. 192a]

Enter Bounce-bigge, Huge-high, Rumble,
Thunderthwart, Bumb-crack, Thumpapace,
Gyants.

Bounc. Come, Enter, Peace is not a trade to thriue at,
Rowse vp your clubs: Huge: Soft let's à while be priuate,
ffor many dayngers rashly haue begunne of warrs,
The safest way, is to assault them vnawares
Scena Quinta. They Consult.

Iupiter, Mars, and Pallas,
aloft.

Iup: Yee knowe right well yee powers, the sonnes of Tytan

233 *sacrific'd*] perhaps *sacrificd'* 236 *oēs.*] . added* 238 *shufflinge*] *in* minims indistinct 239 *scufflinge*] ^{2}*f* written over *l* 242 *ne'[e]re'*] deletion and apostrophes added* 243 *They'l'*] *sic* 248 *Bounce-bigge*] altered from *Bounce-bige** 249 *Thumpapace*] perhaps *Thump a pace*

Make head agaynst vs, and they sweare, they'l fight on
These lofty battlements, till they pull vs downe
ffrom our seate royall heauen, and our crowne,
Impale the temples of theyr cursed head,
And each vppon our conquered helmetts tread;
Let vs consult, since there is such occasion,
How to withstand the gyants hot invasion.

Mars Is not our angry looke enough to terrifie
Those meacocks? I will put them in such feare, if I
Come to them once, that they shall aske vs pardon,
vppon
Vppon theyr knees, the like was neuer heard on. [Fol. 192b]
Neuer was kingly lion bearded soe, by ants,
And wilt thou thus be brau'd by silly giants
Revenge it Ioue: Pal: And let them perish all.

Iup: If they persist, assure yee, they shall fall.

Surgunt Gyg*antes*
a consilio.

Bounc. I haue't, lead on, my head hath hatcht' a stratagem,
Shall make the gods, say mortalls once haue matched vm.

They Whisper againe.

Thump It must be soe of force, now let's beginne
To start them from theyr holes. whop, whos' within?
Why Saturne, Iupiter, Mars, Pallas, Mercury,
Come forth from out the holes wherein you lurke or I,
Will soe bethrume yee. vayne tis to resist,
For wee will fetch yee downe, and this same fist,
Shall like a massy club, light on your sculls,
And dinge yee downe, whose hee that disanulls
What wee command: Iup: That doth Olympick Ioue,

265 *withstand*] perhaps *with stand* 267 *meacocks?*] *?* altered from ,* 271 *brau'd*] perhaps *braud'* *giants*] *ts* heavily inked 274 *Gyg*antes] *ga* altered from *an*; ntes added in italic* 275 a consilio] ? written in a similar but smaller hand; ink is slightly faded; perhaps added later 278 *They*] added* *againe.*] added* 280 *within?*] *?* altered from ,*

That swayes the heauens, and at whose nod doth moue
Theyr weightey frame; borne vp by sturdy Atlas,
Doe.
Doe peeuish mortalls thinke for to out pratle vs? [FOL. 193a]
Pal: What tongue was that that breathed forth those menaces?
Boun: I spake them in theyr names. Iup: Did you? are men asses
To runne vppon theyr deaths like madde folke franticke?
Your rashnes moue's vs, if yee perish thanke it.
Mars How now my masters, what are wee in your eyes
That thus with taunts yee should ⟨ ⟩prouocke our furies?
Iup: If they goe forwards as they make prouision,
I heere protest, for, I am noe precisian,
That if this hand doe once but sett aganst 'vm
They all shall die, nothinge shall pay theyr ransome
Thund: Alas did'st neuer heare, minarum stripitus
Is nothinge else, but asinorum crepitus,
And do'st thou thinke to feare vs w^{t}h a fewe,
Of bugg-beare words, and rantinge speeches? mewe!
Pall: Leaue armes poore wretches, tak't frome me yee want age
To fight with vs, vnlesse vppon aduantage.
Rumb: Then neuer trust me, but least you should say,
Wee tooke yee at aduantage, wee'l away,
Vntill such time yee canne prouide your power,
Then.
Then yee shall see wee'l meete ye at an houre: [FOL. 193b]
How say yee sirs to this, are yee content?
For I'le doe nothinge without your consent.
Oẽs Wee all agree: Thump. Gram̃eries, then lett's on.
Bounc. Vppon advantage wee will cope with none
Exeunt Gygantes.

291 *menaces?*] *?* altered from ,* 293 *franticke?*] *?* altered from ,* 294 *moue's*] *sic*
301 *stripitus*] ^{1}*i* altered from *e** 304 *bugg-beare*] ^{2}*g* added* *speeches?*] *?* altered from ,*
mewe!] *!* altered from .* 309 *power*] *r* badly formed 313 *Gram̃eries*] tilde added*
314 *Bounc.*] added*

Iupi: My ribs will burst with anger, fetch à Cooper,
ffor I shall burst except I haue à hoope heere,
What? am I Ioue? how then [c⟨ . ⟩t] can I endure
These haughty speeches? and of men? be sure,
I'le pay them home at lenght, and they shall see,
Ioue's power extends beyond mortallitie.

Exeunt Oēs.

Scena Sexta.

Enter Neptune Pluto Mercury

Nept: Aboute some weighty matters; alls not well,
What is the matter Mercury? canst tell?
Mer: Truly not I, except it bee concerninge
The worlds o'rethrowe, there's somewhat for your learninge.
Plut: Beware Ioue's foes then, if I goe to Acheron
ffor more supply, Ile bringe them that shall make ye runn⟨e.⟩
Ile fetch Alecto, Tysyphon, Mægera,

Rather.

Rather then fayle, but come shalls to this geere ha. [Fol. 194a]
Nep: Wee'l be reuendgd I sweare by this same trident
As rusty as a panne with bacon fri'd in't
Or wee will runne and fetch our mighty Dolphin
With other fishes that are naught but all finne.
Mer: But hee expects your presence shall wee goe?
Plu: Lead on the way weel followe though it snow.

Exeunt:

Chorus.

Some man will aske me, why I should intend
To let them fall which vse me as theyr frend,
Gyants and clubs are seldome seene asunder,
Why should I let them haue the worst you'l wonder.

318 *can*] interlined above deletion with caret* 319 *speeches?*] *?* added* *men?*] *?* altered from ,* 324 interlined between ll. 323 and 325 (*y* of *Mercury* extends over *w* of *well*, l. 325, establishing that l. 324 was added later, although not in a different ink) 326 *Mercury?*] *?* added* *tell?*] *?* added* 342 *frend*] *r* added, with caret

O sirs. I am a club, but yet the knaue,
Put that, and that, togeather, and yee haue
Your answere, I doe loue for to speake playne I,
Hee that's a knaue, will play the knaue with any.
Sedet.

Scena Septima [FOL. 194b]

Enter Iupiter, Mars, Pallas.

Iup: Are not wee kinge of heauen? and sole commander
Of all things in the earth? commande a stand there.
Heere weel take councel what shall best befitt,
The daynger of these times, here each one sitt.
You are not ignorant yee powers of this,
How the incensed giants led amisse,
By over daringe pride, are vp in armes,
knockinge at heauens gates with loud alarmes.
In vayne may mortalls seeke for helpe from me,
What expectation canne they haue to see,
With sacred odours on our altars throwne,
One venge theyr wrongs, that cannot wreake his done.
Whilst wee securely snort and thinke on naught,
The gyants bandy weapons at our throate.
What is your counsell in this case yee powers,
We must doe somethinge in these fewe short howers.

Scena Octaua. Enter Pluto Neptune
Mercury
welcome

Welcome vnto vs brothers, yee haue heard, [FOL. 195a]
How wee were well nigh pulled by the beard
By proud insultinge rebells, that defie,
The power of Ioue, and his æternitie.

352 *heauen?*] ? added* 353 *there.*] . added* 364 *Whilst*] over an erased word, the last letter of which is *y* 366 *counsell*] *s* altered from *c* 367 *somethinge*] perhaps *some thinge* 368–69 *Enter Pluto Neptune / Mercury*] broader pen stroke and crowded lineation 373 *æternitie*] *n* added, with caret*

And now without your ayd, I shall be brought
To narrowe exigents, our crowne beeinge sought.
Nep: What may those rebells be, are wee not Vnus,
Iupiter, Pluto, and Sea-god Neptunus?
Lett vs but knowe them, and if any power
Bee in the Sea's vaste wombe, it shall deuoure
Those rascalls vp/: Plut: The like doth promise Pluto
Brother you need not feare, I'm sure you know,
What power belongs vnto the stigian Prince,
Rest's all at your commande, and hath e're since
Wee first enioyd the crowne of Erebus,
And were the master of three throated Cerberus,
Wee'l send our furies with theyr locks soe snakie,
And looke to't then, (whoe e're yee be) they'l make yee,
Repent the wrongs that yee haue done our brother,
Twere better yee had medled with some other.
I make.
Iup: I make noe question, for I always finde [Fol. 195b]
That yee are ever vnto me thus kinde.
Pallas, relate vnto them whoe they be
That doe molest vs, for it vexeth me
To thinke vppon them: Pal: They are Titan's issue,
And sweare they will haue Ioue, howe're they misse you,
Vnlesse his crowne Ioue wil to them resigne,
They'l pull vs all from heauen: Plut: This is fine.
Iup: ffirst wee wil sende our Herauld Mercury,
To summon them to parlie ere they die,
If they refuse our proferd amitie,
Theres none but die's, or else I, am not I.
Goe Mercury, and vnto every Gyant,
Say for what's [is] past, to pardon wee are plyant,
If they will leaue of this hostilitie,

376 *rebells*] *s* heavily inked 383 *Rest's*] *sic* 391 *kinde*] *in* minims indistinct
401 *die's*] *sic* 403 *for*] interlinear addition* *what's*] *'s* added at time of *is* deletion*

And part in peace: if not, wee haue abilitie,
To make them vayle the top flags of theyr pride,
This tel them Mercury. ——— Exit Mer:
Our power is tri'd',
Thinke not yee powers, that it [was] is feare in vs,
That makes vs sue for peace vnto them thus;
noe.
Noe, 'tis because wee pitty theyr estate, [FOL. 196a]
And would reclayme them 'fore it were to late,
But whoe canne helpe it, they must fall i'th'end,
That with a greater then themselues contend.
Exeunt Oēs

Scena Nona

Enter Bounce bige, Thunderthwart, Huge-high
Rumble, Bumb-cracke, Thumpapace
Rounciuall.

Bounc. Wee sent for you greate Rounciuall, to ayde vs
Agaynst the gods, because they doe vpbrayd vs,
That wee haue ne'[e]re' a woman to meet pallas,
Now thou shalt bee her, wilt thou not?: Rounc. Alas;
Thinke yee that Rumbles wife, doth care a figge
To fight alone with all, noe, noe, Bounce bigge
Ile with one little finger, and a thumbe take,
Three gyants, Huge-high, Thunderthwart, and Bumb-cracke
And beate them all at once: Bounc: Yet threat not Rounciuall,
Thoul't [wilt] haue to much to doe at once of all
Vs three to fight with, fight with her and spare not
And medle not with vs: Rounc: D'yee thinke I dare not,
Or doe.
Or doe yee thinke this arme, this brawny arme, [FOL. 196b]

408 *tri'd'*] *sic* 409 *is*] interlined above deletion *vs*] *v* written over another letter, or badly inked 412 *them*] interlinear addition with caret* 413 *end*,] perhaps *end*. 417 *Bounce bige*] perhaps *Bounce-bige* 422 *ne'[e]re'*] *e* deleted* 423 *not?:*] *sic; ?* added 424 *Rumbles*] *u* altered from *a* 425 *bigge*] altered from *bige*,* 429 *Thou'lt*] *'lt* added at time of deletion

If it hitt right, cannot doe any harme,?
Wel though I speake not, I wil warrant more,
When't come's to th' push, else Im an arrant whore.
Bounc. It is enough, letts stay noe more Ioue's pleasure,
Shall wee be fooles to wayte vppon his leasure?
Then wee shall make wise worke, tis time we were,
(If you will credit me) aboute this geere.
I vowe, ere night that I will haue Diana
To be my spouse, my masters, what doe yee meane? Why na na
When shall's about it?: Huge: Instantly, and Vulcan
you shall haue hornes enough, looke that your scull canne
Hould out you'r best.—— Enter Mercury
Thund: O are yee come at lenght,
How now, ha's Iupiter prepard his strenght,
What answere sends he, shall wee to the fight?
Mer: He send's these words by me, to wish each on
To fart in peace, and he on that condition
Wil pardon what is past, but if he see,
you wil not leaue of this hostilitie,
And.
And parte in peace, he wil pull downe your pride. [FOL. 197a]
Bounc Gramercies Iupiter, that shall be tri'd'.
Rum: Tell Iupiter from[e] vs, wee'l ne'[e]re' lay downe
These armes we beare, till wee haue crackt his crowne.
Soe gett yee gone, and tel him wee are comminge,
With proud defiance, though wee haue noe drumminge.
Exit Mer:
Huge Come on my masters now we all must stand to't,
Thump There's not à giant, but must sett a hand to't.

435 *come's*] *sic* 441 *na na*] perhaps *nana* 442 *it?:*] *sic*; *?* added in different ink; no attempt made to delete *:* 443 *you*] perhaps *You* 446 *ha's*] *sic* 447 *fight?*] *?* added* 448 *send's*] *sic* 451 *you*] perhaps *You* 453 *tri'd'*] *sic* 454 *from[e]*] *e* deleted* 455 *crackt*] ^{2}*c* added* 456 *yee*] ^{2}*e* added* 460 *Thump*] added*

Bumb: Set/ forwards then, yonder next night wee'l suppe,
And Ioue's owne Ganimed shall fill our cuppe.
Exeunt Oēs.

Scena Decima.
Thunder.
Enter Iupiter, Mars, Pallas, Neptune,
Pluto, in armes.

Iup: Thus soare agaynst our wil, wee are constrayd,
To vse these weapons, yet if they refraynd:
Now at the lenght, why, wee would be content
To let them liue still: Mercury was sent
To shew our minde vnto them, wee shall knowe
When hee's return'd if they'l submit or noe.

Pal: And heere he come's, now Mercury what newes? Enter Mer:
What.
What say the giants? doe they yet refuse [FOL. 197b]
To be in league and amitie? :Mer: They doe.
And thus they bade me I should say to you;
Tel Iupiter (say they) wee'l ne'er lay downe
These armes wee beare, til wee haue crackt his crowne.
That's all that they would send thee for an answere,
And they haue promis't to be heere anon sir.

Nep. And heere they are indeed,——
Enter Bounce bigge, Hughigh, Rumble,
Rounciuall, Thunderthwart, Bumbcracke,
Thumpapace, every one w^th^
his hill.

Bounc. My masters, Come,
Tis à fowle fault that wee should want a drumme.
But tis noe matter, Bumbcracke, thou and Rumble,

461 /] added* to separate *Set* and *forwards* 468 *soare*] *a* added, with caret* *constrayd*] *sic* 474 *come's*] *sic* 475 speech prefix for Iupiter is missing *giants?*] *?* added* 480 *that*] interlinear addition with caret* 483 *Bounce bigge*] perhaps *Bounce-bigge* *bigge*,] altered from *bige** 489 *matter*] *tt* heavily inked and closely spaced

Doe but your heads, and tayles, togeather iumble,
And they shall heare soe hydeous a larume,
That had wee naught else, that alone would feare''vm.
Thunderthwart, thou and I, wee must not blunder,
But put our best voyce to't, t'out cracke theyr thunder.
You must lay load on Thump-apace; Hughigh, thou,
Our.
Our t'allest giant, and wee know, noe Cowe, [Fol. 198a]
O wert thou, O, O, but three horsloaues higher,
Thou shouldst throw 'mongst them Ætna all on fier.
Come courage harts, now giue the signall Rumble,
Rounc: Not hee good Bouncebigge, he, he. can but fumble,
Let Bumbcracke do't, he hath the shriller trumpet,
Then all at heauen braue gogs, vp, vp and thump it.
Bounc. Pelion lie thou there, whilst I gett vp on thee,
Now I am well, make haste, you'l come anon yee?
Hugh. At hand, at hand, and whist I brauely rush on,
Lie still greate Ossa, thou must be my cushion.
Thund Heere Malagolum stand thou noble hillo,
Ossa's his cushion, thou must be my pillowe.
Bumb: Vp looby, vp, and now that I see him on,
Let Bumbcracke roare on thy top plumpe Plymlimmon.
Rumb: Olympus, heere Rumble thus bestrides thee,
Wince not Olympus, cast not him that rides thee.
Thump: Come Rounciuall, we're lag, but weel' not hammer,
Wee'le chuse those hills, that lie besides the Grammer.
Heteroclits, that are in mappe of Lillie,
In that his treatise, where he is soe hillie,
Ile take dindĩmus, and take thou Gargãrus,
Now wee are all fitted, to't Ioue, never spare vs. [Fol. 198b]

491 *heare*] *a* altered from *e** *a larume*] perhaps *alarume* 492 *feare''vm*] *sic* 496 *t'allest*] *sic* 498 *shouldst*] *st* added* 500 *Bouncebigge*] altered from *Bouncebige** 503 *vp on*] perhaps *vpon* *thee,*] ^{2}*e* added* 504 *yee?*] *?* added* 505 *whist*] *sic* 509 *Bumb*] altered from *Rum** 513 *we're*] ^{1}*e* altered from *a** 515 *mappe*] altered from *mape** *Lillie,*] *L* altered from *l** 516 *In*] *n* written over another letter 517 *dindĩmus*] tilde added* *Gargãrus*] tilde added* 517 catchword lost in cropping

Iup: Stand still rebellious race, and come no nigher,
There shall not any put a foote vp higher.
What madnes makes yee, that yee should affect
The regall throne of heauen? doe wee neglect
[The regall throne of heauen]
To mannage our affayres, as doth require
The seate wee now possesse? that yee aspire,
To sit in our tribunall, once more wee'l proffer
Calme peace vnto yee, if yee refuse our offer,
Henceforth with thunder armd, we will transfixe
Those hatefull bulkes of yours, I sweare by stix.

Bounc. Twittle come twats, nor I, nor thunderthwart heere,
Will ere giue or'e, til thy guttes be our garter:

Thump: Come giue the onset and whilst wee doe mount heere,
Good honest Bouncebigg, giue the first encounter.

Mars Degenerate earths-sonnes, how dare ye aduenture
To breake the præscript limitts of the center
To which yee were confin'd? was heauen alotted
To earth borne bratts? I thinke yee are besotted.
Vnlike are yee vnto the earth belowe,
ffrome
ffrom whence yee had your beeinge, yee shall knowe [FOL. 199a]
Heauen's not a place for any of the brood
Of that grosse Element: Thund: Wel sir, twere good
You held your tongue, on noble Clim a the clough,
They shall perceiue wee come not from[e] the Plough.

Oēs Set forwards then: Bounc: Come on my lads of mettle,
There's not à god, but we'el be sure to nettle.
fight

522 *heauen?*] ? added* 525 *possesse?*] ? altered from ,* 526 *proffer*] *pr* interlinear addition with caret* 541 *Wel*] perhaps *wel* 542 *You*] perhaps *you* 543 *from[e]*] *e* deleted*

Iup: Nay then Ile send for ayde and strayte there wool come
One that will make yee shake, arise trisulcum
Enter Thunderboult
Kill Bounc bige first then take them all in order,
That were the causers of this greate disorder,
Die rogues and rascalls, now that yee are vnderhoult
Some stifled with the hills some with our thunderboult
Throwes downe the gyants
Leaue thunderboult hie thee to Vulcan Smith,
New edge and mettle to repayre thee with.
Exit Thunderboult.
the Gyants all wounded
Bounc. Hould, hould, greate Ioue, thou art able for to thwart all
Our fond proceedings, I am but a mortall,
ffor now I wel perceiue, my men are all
Put to the worst, and I, that am theyr generall
haue.

547–58] text of cancel pasted on (as flap) Fol. 199a:
Nay, since I am thus vrgde', Ile' call one w'ul' come
Shall peps them all at once, Fulmen trisulcũ .
Enter Thunderbolt in à furies coate, ougly vizard,
and à wheele of crackers.
Goe herald of our wrath, now th' are vnder holt
With thy Cacofurgoes, dispatch Thunderbolt.
Kill that same Bouncebig, kill them all pell=mell
Tumble each giant from his hill to hell.
Thunderbolt lets flie, the giants droop on the
tops of their hills.
Die rogues and rascalls, our fell anger vnder
Some pressd' with hills, some knockd' downe wth our thunder.
Thunderbolt ha' done, hie thee to Vulcan Smith
New edge and mettaile to repaire thee with.
The giants are slaine, and carried out on the
backes of ye hills.
Bouncebig wounded on the top of his hill, speakes.

1 *w'ul'*] *sic* 2 *peps*] *sic* 6 *Cacofurgoes*] *r* imperfectly formed; altered from *e* 12 *our*] interlinear addition with caret 14 *repaire*] *ir* written very closely together; perhaps *repare*

547 *wool*] ^{2}*o* very slightly formed; perhaps *wol* 550 *Bounc bige*] perhaps *Bouncbige* 555 *hie*] altered from *high*; *e* written over an erased *g* and *h* erased 560 *fond*] *d* resembles *e*

Haue my deaths wounde, my bloud beginnes to issue, [Fol. 199b]
And what wee doe, you care not fort' a' rish you.
Stoope Pelion, thou that wast my stoute vphoulder,
Let me come downe good Pelion, from thy shoulder,
ffor I must goe vnto the sculler Charon,
Whoe churlish chuffe, will scarce agree to beare on
Ouer to stix, vnlesse he pay à shillinge,
Thus the whole world liues only vppon pillinge
But O I die, à boate, à boate, good ferriman,
To pluto Hoe, farewel, I knowe there's ne'[e]r'aman,
But wel can witnesse, not to tel a lie a'nt,
That I heere fall like a couragious Giant.
Moritur
Nep: Downe downe insultinge rebels, on your helmes
Neptune triumphāt treads: Iup: And Ioue or'ewhelmes
the mountaynes on you: Plut: Eearth and Bouncbig's helmet,
now they doe kisse each other, they are wel mett.
Iup: Thus are wee reinvested and posest,
Of heauens seate royall, neuer shall wee rest,
And perfectly heauens potent kinge be crownd,
Til (as our oath requires) the world be drownd,
And
And purgd' from all theese miscreants, and then [Fol. 200a]
We shall effect our full revendge on men./
In the meane time, brothers to you wee owe
Our loue for your assistance, and we'el shewe
Heereafter how wee prise yee. To you all
Ioue giues his generall thankes. Now letts in, and reuel,
Since that the hills and mountaynes are layd leuel.
Exeunt Oēs.

564 *fort' a' rish*] *sic* 572 *ne'[e]r'aman*] *e'* altered from *ee** 573 *a'nt*] *a* altered from *o**, for rhyme with *Giant* (l. 574) 576 *downe*] interlinear addition with caret* 578 *Eearth*] *sic* 578–9 *the . . . now*] *sic* 580 *posest*] descender of ^{1}*s* is split; scribe attempted revision to *possest* 584 *purgd'*] *gd'* altered from *dg** 588 *yee. To*] altered from *yee to** *you*] *u* has an extra minim; perhaps *yow*

Epilogus

What? sit yee still sirs? what is it kinde freends
That yee expect more? for the play heere ends,
I canne assure yee. O then I suppose,
Yee doe expect some ceremonious close,
And pleasinge vpshot, if 't bee soe, you'l finde
Your expectations frustrate, since our minde,
Was (as in all the progresse) soe to make
A mocke=play in the end, yet for your sake
Thus much I'le say, they that Ioues' power withstand,
Humbly submitt[s] [himselfe] themselues vnto your hand[s]/./.

ffinis./

593 *kinde*] *in* minims indistinct 594 *more?*] *?* added* 602 *themselues*] above deletion* *hand[s]*] *s* deleted*

FOLGER MANUSCRIPT J.a.2

I. PROVENANCE AND DATE

MANUSCRIPT J.a.2 (at one time catalogued in the Folger as Manuscript 2203.2) was obtained from Maggs Brothers in 1933.[1] The manuscript had been described in the Sotheby's auction catalogue for 3–7 June 1929 as a 'Commonplace Book, Manuscript on Paper . . . 17th century.'[2] This catalogue does not describe the manuscript's provenance, about which nothing further is known.

The name 'Fra. Corbet' appears on the third leaf of the first gathering (0.3a), which is the manuscript's flyleaf. A Francis Corbet was admitted as a pensioner at Caius College, Cambridge, on 26 October 1584, and matriculated that year. He received his BA in 1587/8.[3] In the Maggs catalogue it was assumed that the manuscript had belonged to this Francis Corbet, probably because he was the elder brother of Clement Corbet, Master of Trinity Hall, Cambridge—the university with which, as will be seen, the manuscript has connections. But the dates of this Francis are too early for an association with J.a.2, and the most likely 'Fra. Corbet' is the one identified by Bowers (p. 124). This one matriculated as a pensioner at St John's College, Cambridge, at Easter 1619. He received his BA in 1622/3 and his MA in 1626,[4] the latest date which can be associated with the manuscript (see below). This Francis Corbet may have been the second son of Thomas Corbet of Yorkshire: and, if so, he became Rector of Patrington in Yorkshire.[5]

On the inverted verso of the second leaf of the final (that is, the twenty-sixth) gathering is written the name 'Alex fFissher'. No Alexander Fisher has been identified with Cambridge connections. An individual with that name was educated at Oxford: he received his BA from Merton in 1617, was elected a Fellow there in 1619, and received his MA in 1623. He was vicar of St Peter's-in-the-East from 1636 to 1646 and also held the title of Burmington.[6] The signature may indicate a former owner of the manuscript and could have been added at any time after the collection of texts had been copied out.

[1] *English Literature and History Fifteenth to the Eighteenth Centuries* (London, 1932) Catalogue 572, item 266, pp. 57–8.

[2] *Catalogue of Valuable Printed Books and Manuscripts* (London, 3–7 June 1929), item 507, p. 80.

[3] *Alumni Cantabrigienses*, i. 396 (work cited on p. 58, n. 2).

[4] Ibid. i. 396.

[5] J. W. Clay (ed.), *Dugdale's Visitation of Yorkshire*, 3 vols. (Exeter, 1917), iii, 442. Another Corbet, a glazier, is mentioned in the Bursar's account at Caius College in 1615–16: see G. C. Moore Smith (ed.), 'The Academic Drama at Cambridge: Extracts from College Records,' Malone Society *Collections II*, pt. ii (Oxford, 1923), 227.

[6] George C. Brodrick, *Memorials of Merton College*, Oxford Historical Society, vol. 4 (Oxford, 1885), 281, 355.

The other annotations, in pencil, appear to be modern. In the lower left-hand corner of the flyleaf there is the mark 'M.v 32', which is presumably an abbreviation for M[aggs] v[olume for 19]32. The note 'Lee:' in the upper left-hand corner is probably a bookseller's or auctioneer's mark: it may indicate a price in code.

Unlike J.a.1, J.a.2 is a single manuscript miscellany. In every place where it is possible to identify the authors or those mentioned in the manuscript—for example, in the cast list for the Latin play *Cancer* (Item 5)—J.a.2 proclaims its association with Cambridge. The two most considerable dramatic pieces in the manuscript, Samuel Brooke's *Melanthe* (Item 3) and the anonymous *Cancer*, were both performed at Trinity College. Theodore Bathurst, author of a Latin version of Spenser's *The Shepheardes Calender* (Item 8), was a Fellow of Pembroke. John Davenant, whose Latin comments on quodlibets form Item 9 and whose theological treatise on grace is Item 15, was President of Queens' College from 1614 to 1621. As mentioned above, the Francis Corbet associated with the manuscript was at St John's in the 1620s. It would appear that one owner in Cambridge—perhaps Corbet—paid scribes over a period of time to copy material he judged interesting or significant. It is equally possible that the collection was copied out, at intervals, for a college or other library and that it then passed into the hands of a private owner.

Although it would be unwise to argue that the contents of J.a.2 were copied in the order of their composition, it is nevertheless true that the latest date which can be associated with the manuscript is that of the final item. This is an order of penance enjoined on one Matthew Hodson, which is dated 3 August 1626. The confusion of the scribes over the correct position for the charges against Roger Brierly (Item 13, 84b–85a, with the title repeated on 88b) may suggest seriatim composition. The earliest date one may be sure about for a text in J.a.2 is 1615, which is when the Trinity play *Melanthe* was acted. Bowers suggests 1610–12 for *Cancer*, but his own evidence appears to contradict this, as it may also Bentley's suggestion of 1611–13. The terminal dates for J.a.2 are, therefore, at the very earliest, 1610, and at the latest 1626—with a greater likelihood for the years 1615–26.

II. CONTENTS

The description of J.a.2 offered by Bowers in 1959 was incomplete and in certain instances incorrect. Further research has enabled the editors to correct some of his identifications. Corrections have also been made to Bowers's bibliographical information, but his numbering of J.a.2's contents has been retained, corrected, and supplemented. The separate pieces are as follows:

1. *A Conflicte between death & youth*. Fols. 1a–2a.
 This is a *contemptus mundi* poem cast in a debate form.
2. An astrological figure. Fol. 2b.
 A figure on the lower half of the page, which indicates 'the influence of the planets on the different hours of the day' (Bowers, p. 125).
3. *Melanthe*. Fols. 3a–24b. [After 1615]
 This play, a Latin pastoral, was written by Samuel Brooke of Trinity College, Cambridge. It was played at Trinity before King James on 10 March 1614/15 and was published in 1615.[1] On Fol. 3a there is a *dramatis personae*, with the names of the Trinity actors, which agrees with another list in manuscript in the British Library and with additions made by hand in the Bodleian copy of the printed edition. In J.a.2, the third leaf of the sixth gathering, containing parts of scenes 3 and 6 and all of scenes 4 and 5 (some 90 lines), is missing.[2]
4. *Ruff, Band, and Cuff*. Fol. 25a–b. [After 1615]
 Discussed below, p. 133.
5. *Cancer*. Fols. 26a–48a.
 This anonymous Latin comedy was adapted from Leonardo Salviati's Italian comedy *Il Granchio* (Florence, 1566 and 1606). A printed version of the Latin text was published in 1648,[3] and this has a different epilogue from the one in J.a.2. On Fol. 26a there is, as with *Melanthe*, a *dramatis personae* and a list of the Trinity College undergraduates who acted in the play. In his analysis of the list of actors, Bowers attempts to identify who played which parts, but the dates he gives for attendances at Cambridge run from 1601 to 1616. The date he conjectures for *Cancer*, 1610–12, is contradicted by the presence of Henry Blaxton, whom he records (p. 127) as being at Cambridge in 1616. This may also make untenable Bentley's date for the play.[4]
6. *Preist the Barbar*, *Sweetball his Man*. Fols. 48b–49a.
 Discussed below, pp. 138–9.
7. *Gowne, Hood, and Capp*. Fols. 49b–50a.
 Discussed below, p. 143.
8. *Kalendarium Pastorale, seu Spenceri Pastor, Romano indutus centenculo*. Fols. 50b–79b.
 A Latin version of Spenser's *The Shepheardes Calender*. The translation is

[1] STC 17800; W. W. Greg, *A Bibliography of the English Printed Drama to the Restoration*, 4 vols. (London, 1939–58), ii. 937–8. The printed version was edited by J. S. G. Bolton, Yale Studies in English 79 (New Haven, 1928).

[2] In Bolton's edition, IV. iii. 27 to IV. vi. 2. For the manuscript in the British Library, MS Add. 6211, p. 33; see Bolton, pp. 202–5.

[3] Wing STC C423A; Greg, ii. 952–3.

[4] G. E. Bentley, v. 1298 (work cited on p. 3, n. 1).

ascribed in the manuscript (Fol. 50b) to 'Magister Batters,' that is, Theodore Bathurst, Fellow of Pembroke College, Cambridge in 1608. Bathurst's translation was published in 1653.[1]

9. Latin comments on the quodlibets. Fol. 80a–b.

An attribution in the margin of the manuscript gives the authors (or speakers) as Dr Dauenett and Mr Brittaine. A John Davenant received his BD from Queens' College, Cambridge, in 1606 and was President of Queens' from 1614 to 1621. A Lawrence Bretton received his BD from Queens' in 1615.

10. *The Parliament Fart*. Fols. 81a–82a.

This is a version of the popular anti-Puritan poem, of which there is a vast number of manuscript texts. Its subject-matter was added to and adapted over the years, to suit changing political circumstances: its doggerel couplets could be deleted, or fresh ones supplied, with ease. One version reached print in Mennis's and Smith's *Musarum Deliciae* (London, 1655), while another was published in Alexander Brome's *Rump* (London, 1662).[2] The lists of persons in J.a.2 and the printed versions overlap only in part. There can be no reference to a specific Parliament in the printed texts since Richard Martin, who appears in both J.a.2 and *Musarum* died in 1618 (having sat in Parliament in 1601 and 1604–11), while Sir Richard Lovelace, who appears in *Musarum*, was born in 1618 and may never have sat in Parliament at all. The J.a.2 version seems to refer to the 1604–11 Parliament; the death in 1616 of Sir Jerome in Folio, that is, Sir Jerome Bowes, provides a *terminus ad quem*.

11. An anti-French political couplet. Fol. 82a.

A jibe at Pierre Cotton (1564–1626), the Jesuit confessor to Henri IV. Bowers transcribes the couplet wrongly and omits the translation which follows it. The text in full is this:

Voulez vou scauojr qui a trouble la France
La Plume, & la cere, La Cotton, & L'Ancre

Who hath troubled Fraunce would you thinke
The Quill, the wax, the Cotton, & ye Inke./

12. *Synopsis Physicae Christianae*. Fols. 82b–84a.

An anonymous text in Latin on the physical and moral universe.

13. *Certayne erronjous proceedings, gathered from ye mouth of Mr. Brierly, & some of his hearers*. Fols. 84b–85a. Dated 1617.

[1] Wing STC S4966, S4968. For an analysis of this printed version, see F. R. Johnson, *A Critical Bibliography of the Works of Edmund Spenser* (Baltimore, 1933), pp. 8–10. Manuscript versions are discussed by Beal, vol. 1, pt. 2, p. 524 (work cited on p. 8, n. 3).

[2] *Musarum Deliciae*: Wing STC M1710, sigs. F1–4; *Rump*: Wing STC B4851, sigs. E7–8.

Roger Brierley was a Puritan preacher who was prosecuted on fifty counts of religious error. At his trial, the case against him was dismissed by Bishop Toby Matthew. The date of the trial is uncertain, but it must have been before 1628, which is when Matthew died. In J.a.2, the list of fifty statements of error (such as 'Thatt faith and feelinge are inseparable') cannot be identical with the original fifty charges, and must date after the trial, since the forty-second statement is that 'ye Bishop of Yorke is a 2^{d} Faelix, for when he was about to ꝓnounce sentence of silencing Mr Brierly there fell such a trembling and quakinge vpon him, y^{t} hee durst nott doe it.' If the date given for J.a.2's list of statements, 1617, is correct, then Brierley's trial may have occurred considerably earlier than 1628.

14. Latin comments on the quodlibets. Fol. 85b.

Attributed to Dr Dauenett in the manuscript (see Item 9).

15. *De Meritis*. Fol. 86a.

A brief Latin theological treatise, attributed in the manuscript to Dr Dauenett.

16a. A scribal fair copy of Sir Walter Ralegh's last letter to his wife. Fols. 86b–87a (ll. 1–13).[1]

16b. 'Euen such is Time wch takes in trust'. Fol. 87a (ll. 14–23).

A scribal fair copy of Ralegh's poem. The text varies from the version printed by Agnes M. C. Latham in her edition of Ralegh's poems, *Poems*.[2]

17. Latin epigram. Fol. 87a (ll. 24–5).

The epigram reads as follows:

> Conde iura coquus, quid ni ? condire peritus
> Jura coquus, sed no condere iura coquus.

Bowers translates these lines as 'Draw up laws, cook, why not? A cook is skilled / In seasoning sauces, not in drawing up laws.' The epigram is one of the many pieces which circulated in 1616 when Sir Edward Coke was deprived of his position as Lord Chief Justice for upholding common law over the rights of King James I. Another version of the epigram is quoted by Catherine Drinker Bowen in her biography of Coke.[3]

18.[4] An order of penance. Fol. 88b. Dated 3 August 1626.

The penance is enjoined on Matthew Hodson of Hitchin. It is signed by 'ja. Rolfe' at Whethamsted.

[1] See *The Works of Sir Walter Ralegh*, 8 vols. (Oxford 1829), viii. 648–50.

[2] London, 1951, p. 72 (Poem XL).

[3] *The Lion and the Throne* (Boston, 1957), p. 390.

[4] Bowers reversed the order of Items 18 and 19. His Item 18 should be the final one (on 88b), and his Item 19 should be the penultimate piece (on 87b–88a).

19.[1] Anonymous Latin theological comments. Fols. 87b–88a.
The comments begin with 'Lu. 19.8.'

20.* Unnumbered in Bowers. A title: 'Certain erronjous proceedings gathered from ye mouth of Mr Brierly, & some of his hearers.' Fol. 88b.
This title was copied in error from Item 13, which suggests some uncertainty on the part of the scribe about the position of the piece.

III. PAPER AND HANDS

Manuscript J.a.2 is far less complicated, bibliographically speaking, than J.a.1. It is a quarto of twenty-six gatherings, six of which are anomalous. In the first gathering, the first leaf, 0.1, is missing; leaf 0.2 has been cropped to 40 mm, then folded in half and stitched; 0.3 contains the 'Fra. Corbet' inscription; and 0.4 is blank. In the second gathering, the first leaf, 0.5, is trimmed. The Folger numbering begins on the second leaf of this gathering. The third leaf of the sixth gathering is missing (see above, under Item 3). The fourth leaf on the twenty-second gathering has been trimmed to 10 mm. It is difficult to determine whether two pages of 'The Parliament Fart', Item 10, have been omitted. One hint of revision or loss may be that the final fourteen lines of the poem are on the first leaf of the twenty-third gathering (Fol. 82a) in a different hand. The twenty-fourth gathering, which is trimmed to 9 mm, lacks its final leaf. All of the twenty-fifth gathering has been trimmed, although the upper margins and some signs of writing are still visible. In the final (that is, the twenty-sixth) gathering, the name 'Alex fFissher' is written on the verso of the second leaf, and the third leaf has been trimmed, folded, and stitched down to 15 mm.

Each page measures approximately 150 × 198 mm. On each recto there is a left margin of approximately 15 mm, and a right margin of 24 mm. These measurements are reversed on each verso. There is a double horizontal margin, 8 mm deep, drawn across the full page, beginning approximately 9 mm from the top.

Only one watermark appears in J.a.2: it resembles most closely the items in Briquet, Armoirie 1380 (1591), Churchill, Coat of Arms 285 (1603), and Heawood, Cockatrice 841 (1621).[2] The chain lines are approximately 25 mm apart. The final gathering is too tightly bound into the sewing for the watermark to be identified.

[1] Bowers reversed the order of Items 18 and 19. His Item 18 should be the final one (on 88b), and his Item 19 should be the penultimate piece (on 87b–88a).

[2] Charles Moise Briquet, *Les Filigranes*, ed. Allan Stevenson (Amsterdam, 1968); Churchill and Heawood (works cited on p. 10, n. 2).

There are sixteen hands at work in J.a.2:

Item number	*Title, or Contents*	*Folio*	*Hand*
1	*A Conflicte*	1a–2a	A
2	Astrological figure	2b	B (italic in the figure, secretary below)
3	*Melanthe*	3a–24b	C[1]
		24b (Epilogue)	D
4	*Ruff, Band, and Cuff*	25a–b	C′ (see Plate 6)
5	*Cancer*	26a–48a	C
6	*Preist the Barbar*	48b–49a	C′ (see Plate 7)
7	*Gowne, Hood, and Capp*	49b–50a	C′
8	*Kalendarium Pastorale*	50b–79b	C (and E; see below)
9	Latin comments	80a–b ll. 1–8	F
9		80b l. 9–end	G
10	*The Parliament Fart*	81a–b	H
10		82a	I
11	Political couplet	82a	I
12	*Synopsis*	82b–84a	J
13	*Certayne erronjous proceedings*	84b–85a	K
14	Latin comments	85b	K
15	*De Meritis*	86a	L
16a	Ralegh letter	86b–87a, ll. 1–13	M
16b	Ralegh poem	87a, ll. 14–23	N
17	Latin epigram	87a, ll. 24–25	N
19	Theological comments	87b–88a	O
20*	Miscopied title	88b	K
18	Order of penance	88b	P

[1] The running title, 'Melanthe', on every page, and the *dramatis personae* on Fol. 3a are in the same hand but a different ink.

The most difficult of these hands to distinguish are C and C′. They appear to be the same, with a different pen encouraging smaller lettering in C′. Hand K is also similar to Hand C. In Item 8, the Latin version of *The Shepheardes Calender*, there are two stages of correction: (i) Hand C returns with a different ink to make corrections and marginal annotations, and (ii) Hand E, using black ink, works on the poem from its beginning (Fol. 50b) through to the end of August (69a). E appears to have been attending to the translation, as he supplies some corrections and line numbers. He also modernizes the question marks.

The three plays which are edited here—*Ruff, Band, and Cuff*, *Preist the Barbar, Sweetball his Man*, and *Gowne, Hood, and Capp*—are all transcribed in hand C′ (see Plates 6 and 7). This is a small, tight, italic hand in which it is particularly difficult to distinguish minims, especially *in* from *ni*. The scribe occasionally uses a secretary form, particularly when writing terminal *d*, and sometimes *the*. He has a tendency to run words together, so that 'a n'ey' or 'top gallants' (*Preist the Barbar*, ll. 36, 37, and 85) may be one or two words. The first *t* of a word ending in double *tt* is rarely crossed. *L*/*l*, *W*/*w*, and *Y*/*y* are easily confused.

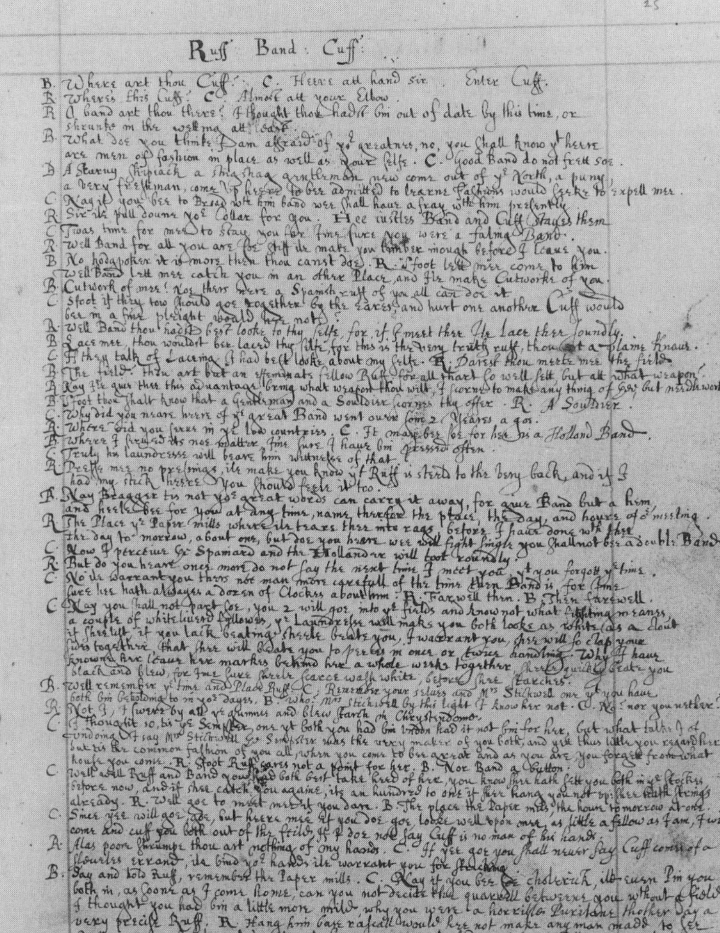

25
Ruff : Band : Cuff :
B. Where art thou Cuff? C. Heere at hand sir. Enter Cuff.
R. Where's this Cuff? C. Almost at your Elbow.
R. A band art thou there? I thought thou hadst bin out of date by this time, or shrunke in the wetting att least.
B. What doe you thinke I am affraid of yor greatnes, no, you shall know yt heere are men of fashion in place as well as your selfe. C. Good Band do not frett soe.
B. A starvy skipiack a shagshag gentleman new come out of ye North, a puny, a very freshman, come up heere to bee admitted to learne fashions would seeke to expell mee.
C. Nay if you bee to Bread wth him band wee shall haue a fray wth him presently.
R. Sir ile pull downe yor Collar for you. Hee iustles Band and Cuff stayes them
C. Twas time for mee to stay you for time sure you were a falling Band.
R. Well Band for all you are soe stiff ile make you limber inough before I leaue you.
B. No hodgpoker it is more then thou canst doe. R. Sfoot lett mee come to him Well Band lett mee catch you in an other Place, and Ile make Cutworke of you.
B. Cutwork of mee! Noe there were a Spanish ruff of you all can doe it.
C. Sfoot if they tow should goe together by the eares, and hurt one another Cuff would bee in a fine pleight would hee not?
R. Well Band thou hadst best looke to thy selfe, for if I meet thee Ile lace thee soundly.
B. Lace mee, thou wouldst bee laced thy selfe, for this is the very truth ruff, thou art a plaine knaue.
C. If they talk of Laceing I had best looke about my selfe. R. Darest thou meete mee the field.
B. The field! thou art but an effeminate fellow Ruff, for all thart so well sett, but at what weapon?
R. Nay Ile giue thee this aduantage bring what weapon thou wilt, I scorne to make any thing of thee but needleworke
B. Sfoot thou shalt know that a Gentleman and a Souldier scornes thy offer. R. A Souldier.
C. Why did you neare heere of ye great Band went ouer some 2 yeares a goe.
R. Where did you serue in ye low countries. C. It may bee soe for hee is a Holland Band.
B. Where I serued its noe matter, time sure I haue bin pressed often.
C. Truly his laundresse will beare him witnesse of that.
R. Presse mee no pressings, ile make you know yt Ruff is steeld to the very back, and if I had my stick heere you should feele it too.
B. Nay Bragger tis not yor great words can carry it away, for giue Band but a hem and heele bee for you at any time, name therefore the place, the day, and houre of or meeting.
R. The Place ye Paper mills where ile teare thee into rags, before I haue done wth thee the day to morrow, about one, but doe you heare wee will fight single you shall not bee a double Band.
C. Now I perceiue ye Spaniard and the Hollander will goe roundly.
R. But do you heare once more do not say the next time I meet you, yt you forgott ye time.
C. No ile warrant you theres not man more carefull of the time then Band is, for sure hee hath alwayes a dozen of Clockes about him. R. Farwell then. B. Then farewell.
C. Nay you shall not part soe, you 2 will goe into ye fields and know not what fighting meanes, a couple of whiteliuerd fellowes, ye Laundresse will make you both looke as white as a clout if shee lift, if you lack beating sheele beate you, I warrant you, shee will so clap your sides together, that shee will beate you to peeces in once or twice handling. Why I haue knowne her leaue her markes behind her a whole weeke together, sheele quickly beate you black and blew, for I me sure sheele scarce wash white, before shee starches.
B. Well remember ye time and Place Ruff. C. Remember your selues and Mrs Stickwell one yt you haue both bin beholding to in yor dayes. B. who? Mrs Stickwell by this light I know her not. C. No? nor you neither?
R. Not I, I sweare by all ye gummes and blew starch in Chrystendome.
C. I thought it so, tis ye Sempster, one yt both you had bin undon had it not bin for her, but what talke I of undoing I say Mrs Stickwell ye Sempster was the very maker of you both, and yet thus little you regard her, but tis the common fashion of you all, when you come to bee great and as you are, you forgett from what house you come. R. Sfoot Ruff cares not a point for her. B. Nor Band a button.
C. Well well Ruff and Band you had both best take heed of her, you know shee hath sett you both in ye stockes before now, and if shee catch you againe, its an hundred to one if shee hang you not up: shee hath strings already. R. Well goe to meet mee if you dare. B. The place the Paper mills, the houre to morrow at one.
C. Since yee will goe, doe, but heere mee if you doe goe, looke well upon mee, as little a fellow as I am, I will come and cuff you both out of the field, if I doe not say Cuff is no man of his hands.
R. Alas poore shrimpe thou art nothing of my hands. C. If yee goe you shall neuer say Cuff come of a flourles errand, ile bind yor hands ile warrant you for stricking.
B. Say and hold Ruff, remember the Paper mills. C. Nay if you bee soe cholerick, ile euen pin you both in, as soone as I come home, can you not decide the quarrell betweene you wthout a fields I thought you had bin a little more milde, why you were a horrible Puritane thother day a very precise Ruff. R. Hang him base rascall would hee not make any man madd to see such a poore snake, yt durst not scarce peepe out of dores befor Collar came to towne now to swagger it soe. C. Come you shallbe freinds Band.
B.

Plate 6: Folger Manuscript J.a.2, Fol. 25a (two-thirds full size): *Ruff, Band, and Cuff*, lines 1 to 64

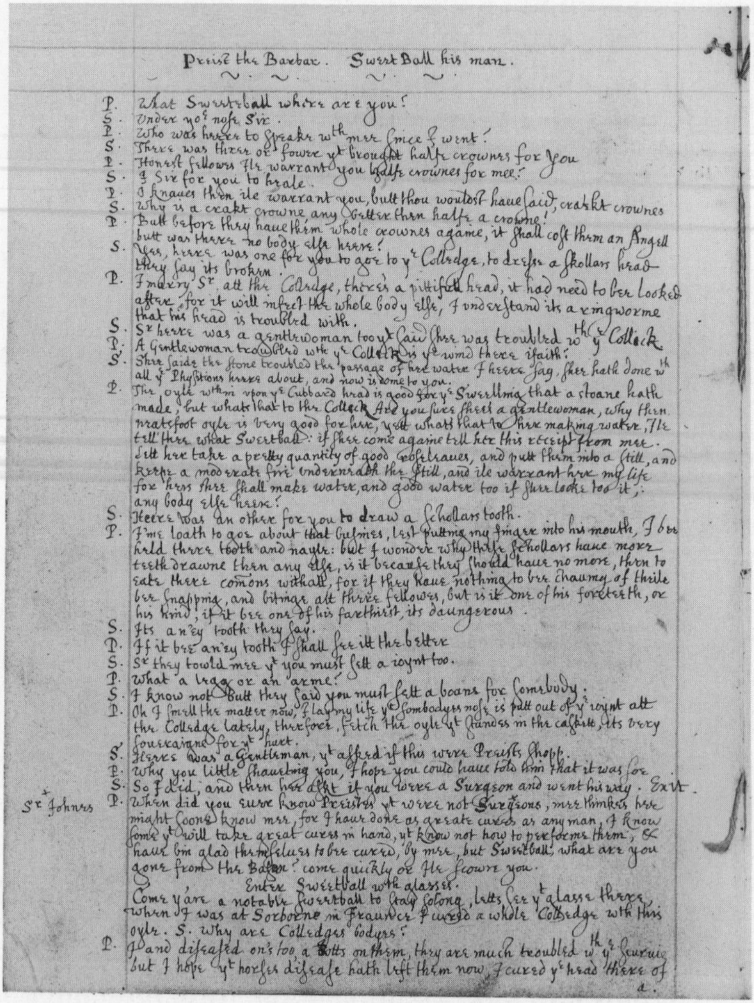

Preist the Barbar. Sweet Ball his man.

P. What Sweeteball where are you?
S. Under yor nose Sir.
P. Who was heere to speake wth mee since I went?
S. There was three or foure yt brought halfe crownes for you
P. Honest fellows Ile warrant you halfe crownes for mee?
S. I Sir for you to heale.
P. I knaues then ile warrant you, butt thou woulds't haue said, crakt crownes
S. Why is a crakt crowne any better then halfe a crowne?
P. Butt before they haue them whole crownes againe, it shall cost them an Angell
butt was there no body else heere?
S. Yes, heere was one for you to goe to ye Colledge, to dresse a skollars head
they say its broken.
P. I marry Sr. att the Colledge, there's a pittifull head, it had need to bee looked
after, for it will infect the whole body else, I understand its a ringworme
that his head is troubled with.
S. Sr heere was a gentlewoman too yt said shee was troubled wth ye Collick.
P. A Gentlewoman troubled wth ye Collick is ye wind there ifaith?
S. Shee saide the stone troubled the passage of her water I heere say, shee hath done wth
all ye Physitions heere about, and now is come to you.
P. The oyle within upon ye Cubbard head is good for ye Swellmg that a stoane hath
made, but whats that to the Collick. Are you sure shee's a gentlewoman, why then
neatsfoot oyle is very good for her, yet whats that to her making water, Ile
tell thee what Sweetball: if shee come againe tell her this receipt from mee.
Lett her take a pretty quantity of good Rose leaues, and putt them into a still, and
keepe a moderate fire underneath the still, and ile warrant her my life
for hers shee shall make water, and good water too if shee looke too it;
any body else heere?
S. Heere was an other for you to draw a schollars tooth.
P. I'me loath to goe about that busines, lest putting my finger into his mouth, I bee
held there tooth and nayle: but I wonder why these schollars haue more
teeth drawne then any else, is it because they should haue no more, then to
eate there comons withall, for if they haue nothing to bee chawing of theile
bee snapping, and bitinge att there fellowes, but is it one of his foreteeth, or
his kind; if it bee one of his farthirst, its daungerous.
S. Its an'ey tooth they say.
P. If it bee an'ey tooth I shall see itt the better
S. Sr they towld mee yt you must sett a ioynt too.
P. What a legg or an arme?
S. I know not butt they said you must sett a boane for somebody:
P. Oh I smell the matter now, I lay my life yt somebodyes nose is putt out of ye ioynt att
the Colledge lately, therefore, fetch the oyle yt standes in the caskitt, its very
soueraigne for yt hurt.
S. Heere was a Gentleman, yt asked if this were Preists shopp.
P. Why you little shaueling you, I hope you could haue told him that it was for
S. So I did, and then hee askt if you were a Surgeon and went his way. Exit
Sr Johns P. When did you euer know Preists yt were not Surgeons, mee thinkes hee
might soone know mee, for I haue done as greate cures as any man, I know
some yt will take great cures in hand, yt know not how to performe them, &
haue bin glad themselues to bee cured, by mee, but Sweetball, what are you
gone from the Barber? come quickly or Ile scowre you.
Enter Sweetball wth glasses.
Come y'are a notable sweetball to stay so long, letts see yt glasse there,
when I was at Sorbonne in Fraunce I cured a whole Colledge wth this
oyle. S. Why are Colledges bodyes?
P. I and diseased on's too, a botts on them, they are much troubled wth ye scurvie
but I hope yt horses disease hath left them now, I cured ye head there of
a.

PLATE 7: FOL. 48b (TWO-THIRDS FULL SIZE): *PREIST THE BARBAR, SWEETBALL HIS MAN*, LINES 1 TO 57

PLAYS FROM J.a.2: TEXTS AND HEADNOTES

Item 4 in J.a.2: Fol. 25a–b *Ruff, Band, and Cuff*

Ruff, Band, and Cuff is transcribed in hand C′ (see Plate 6 and above, p. 130). There are seven textual witnesses to this play, four in manuscript, and three in early printed editions. In addition to the J.a.2 text, the manuscript versions are in the British Library (MS Add. 23723, Fols. 1–3), in the Bodleian (Lyell MS), and among the holdings of the West Yorkshire Archive Service in Bradford (the Hopkinson MS, formerly located at Ashton Hall in Yorkshire). At the head of the Hopkinson MS the 'Comedye' is said to have been 'Acted at Oxford, Feb 24 Anno. Dm. 1646.' The printed editions are all quartos.[1] On the title-page of Q1 (1615), the 'Merrie Dialogve' is said to have been '*Lately acted in a shew in the famous* / Vniversitie of CAMBRIDGE.'[2] Q2 (also 1615) advertises itself on the title-page as 'EXCHANGE / WARE AT THE / SECOND HAND / . . . *lately out, and now newly* / dearned vp.' Q3 appeared in 1661.[3] The variations in phrasing between the seven texts are not substantial. It is unclear which extant text, if any, served as a source for the remainder. Indeed, it is unlikely that any surviving version is the first.

Although *Ruff, Band, and Cuff* does not have as many academic allusions as *Preist the Barbar, Sweetball his Man* and *Gowne, Hood, and Capp*, the play does mention 'a Scuruy shipiack . . . / a very freshman' (ll. 8–9). It is short, plays extensively on words, has no female roles, and needs no special equipment for production. Its implied audience is addressed in line 103 as 'gentlemen'.

Hanford suggested that *Ruff, Band, and Cuff*, together with *Worke for Cutlers. Or, a Merry Dialogue betweene Sword, Rapier, and Dagger*, and *Wine, Beere, Ale & Tobacco*, were all the work of 'the same excellent Cambridge wit'. Moore Smith attributed the play, with understandable doubts, to Thomas Tomkis of Trinity College.[4]

[1] STC 1356–57, Wing M1857; Greg, i. 466–8 (work cited on p. 125, n. 1).

[2] Q1 was reprinted in volume I of *The Old Book Collector's Miscellany*, ed. Charles Hindley (London, 1871); the reprint was reissued in *A Collection of Readable Reprints*, Item 5 (London, 1876).

[3] It was reprinted in 1813 as volume x of *The Harleian Miscellany*, ed. Thomas Park. The Prologue from the Hopkinson manuscript was added, together with 'preferable' readings from that manuscript. It was reprinted again by James Halliwell-Phillipps in *Contributions to Early English Literature* (London, 1849).

[4] James Holly Hanford, 'The Debate Element in Elizabethan Drama,' *Kittredge Anniversary Papers* (Boston, 1913), p. 456; and G. C. Moore Smith, *College Plays Performed in the University of Cambridge* (Cambridge, 1923), p. 8.

Ruff: Band: Cuff: [FOL. 25a]

B. Where art thou Cuff? C. Heere att hand sir. Enter Cuff.

R. Where's this Cuff? C. Almost att your Elbow.

R. O band art thou there? I thought thou hadst bin out of date by this time, or shrunke in the wetting att least.

B. What doe you thinke I am affraid of yor greatnes, no, you shall know y^{t} heere are men of fashion in place as well as your selfe. C. Good Band do not frett soe.

B. A Scuruy shipiack a shigshag gentleman new come out of ye North, a puny, a very freshman, come vp heere to bee admitted to learne Fashions would seeke to expell mee.

C. Nay if you bee to Broad wth him band wee shall haue a fray wth him presently.

R. Sir ile pull doune yor Collar for you. Hee iustles Band and Cuff stayes them

C. Twas time for mee to stay you for Ime sure you were a faling Band.

R. Well Band for all you are soe stiff ile make you limber inough before I leaue you.

B No hodgpoker it is more then thou canst doe. R. 'Sfoot lett mee come to him Well Band lett mee catch you in an other Place, and Ile make Cutworke of you.

B. Cutwork of mee! Noe thers nere a Spanish ruff of you all can doe it

C. Sfoot if they tow should goe together by the eares, and hurt one another Cuff would bee in a fine pleight would hee not?

R. Well Band thou hadst best looke to thy selfe, for if I meet thee Ile lace thee soundly.

B. Lace mee, thou wouldst bee laced thy selfe, for this is the very truth ruff, thou ⟨ar⟩t a plaine Knaue.

C. If they talk of Laceing I had best looke about my selfe. R. Darest thou meete mee the field

B. The field? thou art but an effeminate fellow Ruff. for all thart so well sett, but att what weapon?

R. Nay Ile giue thee this aduantage bring what weapon thou wilt, I scorne to make any thing of thee, but needleworke

B. 'Sfoot thou shalt know that a Gentleman and a Souldier scornes thy offer. R. A Souldier!

C. Why did you neare heere of ye great Band went ouer som 2 Yeares a goe.

R. Where did you serue in ye low countries. C. It may bee soe for hee is a Holland Band.

2 *sir*] perhaps *Sir* 6, 8 *B.*] *B* altered from *R* 8 *Scuruy*] *c* altered from *t* *shigshag*] perhaps *shig shag* 21 *mee the*] *sic*

B. Where I serued its noe matter, I'me sure I haue bin pressed often.

C. Truly his laundresse will beare him wittnesse of that.

R. Presse mee no pressings, ile make you know yt Ruff is steeld to the very back, and if I had my stick heere you should feele it too.

B. Nay Bragger tis not yor great words can carry it away, for giue Band but a hem, and heele bee for you at any time, name therfor the place, the day, and houre of o^{r} meeting.

R. The Place ye Paper mills where ile teare thee into rags, before I haue done wth thee, the day to morrow, about one, but doe you heare wee will fight single you shall not bee a double Band.

C. Now I perceiue the Spaniard and the Hollander will toot roundly.

R. But do you heare once more do not say the next time I meet you, y^{t} you forgott ye time.

C. No ile warrant you thers noe man more carefull of the time then Band is, for Ime sure hee hath alwayes a dozen of Clockes about him. R. Far well then. B. Then farewell.

C. Nay you shall not part soe, you 2 will goe into ye fields and know not what fighting meanes, a couple of white liuerd fellowes, ye Laundresse will make you both looke as white as a clout if she list, if you lack beating sheele beate you, I warrant you, shee will so clap your sides together, that shee will beate you to peeces in once or twice handling. Why I haue knowne her leaue her markes behind her a whole weeke together, sheele quickly beate you black and blew, for Ime sure sheele scarce wash white, before shee starches.

B. Well remember ye time and Place Ruff. C. Remember your selues and Mrs Stichwell one y^{t} you haue both bin beholding to in yor dayes. B. who? Mrs Stitchwell by this light I know her not. C. No? nor you nether?

R. Not I, I swere by all ye gumme and blew starch in Chrystendome.

C. I thought so, tis ye Sempster, one y^{t} both you had bin vndon had it not bin for her, but what talke I of

34 *heare*] *a* altered from *e* 35 *Spaniard*] perhaps *Spainard* 39 *fighting*] g^{1} altered from *t*
40 *white liuerd*] perhaps *whiteliuerd*

vndoing I say Mrs Stichwell the Sempster was the very maker of you both, and yett thus little you regard her, but tis the common fashion of you all, when you come to bee great and as you are, you forget from what house you come. R. Sfoot Ruff cares not a point for her. B. Nor Band a button.

C. Well well Ruff and Band you had both best take heed of her, you know shee hath sett you both in ye stockes before now, and if shee catch you againe, its an hundred to one if shee hang you not vp: shee hath strings already. R. Well goe to meet mee if you dare. B. The place the Paper mils, the houre to morrow at one.

C. Since yee will goe, goe, but heere mee if you doe goe, looke well vpon mee, as little a fellow as I am, I will come and cuff you both out of the feild. If I doe not say Cuff is no man of his hands.

R. Alas poore shrumpe thou art nothing of my hands. C. If yee goe you shall neuer say Cuff comes of a sleueles errand, ile bind yor hands ile warrant you for stricking.

B. Say and hold Ruff, remember the Paper mills. C. Nay if you bee soe cholerick, ile euen Pin you both in, as soone as I come home, can you not decide the quarrell betweene you wthout a field I thought you had bin a little more mild, why you were a horrible Puritane thother day a very precise Ruff. R. Hang him base rascall would hee not make any man madd to see such a poore snake, y^{t} durst not scarce peepe out of dores befor Collar came to towne now to swagger it soe. C. Come you shalbe freinds Band.

B.

B. Freinds wth him. such a base rascall hees a very thredbare fellow I scorne but my man Collar [FOL. 25b] shall goe better euery day in ye weeke then hee, & bee frends wth him?

R. Thy man Collar thy M^{r} thou wouldst say for Ime sure hees thy vp holder.

C. Nay surely hee is his M^{r} or at least his maker, for Bands makes Rags, Rags makes Paper

50 *forget*] perhaps *forgett* 52 ^{1}*you*] perhaps *your* 54 *to morrow*] perhaps *tomorrow* 56 *you*] *y* altered from *c* 67 *vp holder*] perhaps *vpholder* 68 *surely*] altered from *surly*

Paper makes Pastboord, and Pastboord makes a Collar, & I think this is a stiff argument y^{t} hee is his Maker and therfore Master.

R. Well bee hee what hee will bee if I catch his Collar ile cutt him into iaggs, lett mee but Claspe him and ile make him for stirring.

C. But you shall not fight, haue you not frends and neibours to end this controuersie for you, but you must fight, and goe into the field, and there cutt the thred of yor liues, wee will haue none of that, com chose you an Vmpire Band, for it shall bee soe.

B. Since you will force mee toot if Ruff bee content I am willing

C. Ruff you shall bee content. R. If I shall I must, lett him name him.

B. If I may name him Ile haue Mr Handkerchiffe.

C. Nay stay there, hees a most filthy sniueling fellow & a notable Lawyer, heele wipe Your nose of all if you put the case to him. What say you to shirt?

R. Hees a filthy shifting knaue and one whom Band hath bin much beholding vnto, they were ioyned together a long time in Frendshipp.

C. Why then goe to Mr Cap the headman of the towne. B. No I deny that hees a very bad iustice, you may haue him wrought on any side for mony.

R. Ile tell you what then weele go to my Lord Corpus himselfe.

B. Hees not in Towne. R. Yes for this day I saw Sock his Cheefe footman.

C. Heers a doe wth you and my L. Corpus. I would you both were hangd about his neck for mee, I think you would then bee quiet. but I see this strife will nere bee ended vnless I bee Moderator betwene you, you know I am equally allied to you both, shall I bee arbitrator betweene you. R. Content B. Content.

C. Well then Ruff shalbe most accepted of among the Clergy: for hee is the grand fellow, although I know the puritans will not greatly care for him, hee hath such a deale of setting, and they loue standing so well. As for you Band you shall bee most of all made of amongst the gallants, although they shall vse Ruff somtimes for a fashion but not otherwise, But how so euer you need not regard ye giddyheaded

69 *this*] perhaps *his* 95 *how so euer*] perhaps *howsoeuer* *giddyheaded*] perhaps *giddy headed*

multitude, lett them do as they list, somtime respecting one thing, somtime another, But when you come to the Councellours wch are men of Law, and know right from wrong, iudging your worthes to bee equall, they shall preferre neyther, but vse the kindnesse of you both, wearing both a Band and a Ruff how say you are yee both pleased. Speake

B
R } Wee are

C. Then goe before to the next Tauerne and Ile follow after, wth a new band of yor frendshipp drawne, wch I hope these gentlemen will seale wth there handes.

Cuff:

The Epilogue.

Claw mee, and ile claw thee ye Prouerbe goes.
Lett it bee true, wch this our Muse heere shoes.
Cuffs graceth hands, Cuffs debters hans remaine
Lett hands Clapp me, and ile Cuffe them againe

98 *worthes*] perhaps *w'orthes*

Item 6 in J.a.2: Fols. 48b–49a ***Preist the Barbar, Sweetball his Man***

Preist the Barbar, *Sweetball his Man*[1] is transcribed in hand C′ (see Plate 7, and above, p. 130). The play is a comic sketch about shaving in college, and it can be identified beyond doubt as a Cambridge piece. Lines 59–61 refer to the sides of 'ye Colledge', which are said to have 'had a Conuulsion, [and] as for ye walls they had the falling sicknesse, [and] the / Windoes, had a great paine in them.' These are clearly allusions to the great play riot of 6–7 February 1611, when the scholars and porters of St John's pulled down the garden wall in front of Trinity College and broke the glass in the college windows. *Preist the Barbar*, *Sweetball his Man* may well have been written in the same year, since it alludes to repairs made to the college after the riot.[2]

[1] Both Bowers (p. 128) and S. Schoenbaum, p. 203 (work cited on p. 3, n. 3) refer to the play simply as *Preist the Barbar*. The title followed here is the one given in the manuscript, which, like the other two pieces in this edition—*Ruff*, *Band*, *and Cuff* and *Gowne*, *Hood*, *and Capp*—names all the speaking parts.

[2] Alan Nelson, private communication to the editors, 15 December 1987. Nelson will present his evidence in full in his forthcoming REED volume on early drama at Cambridge. See also

Of the several men named Preist who may be linked to the play, the most likely appears to be one Henry Preist, a successful barber and scholar's servant in Cambridge in the early decades of the seventeenth century.[1] A contemporary annotation of line 47 ('When did you euer know Preistes y^{t} were not Surgeons') notices the pun on priests being 'Sr Johnes' or surgeons.[2]

The play is short, has no female roles and few if any requirements for production: everything about it indicates that it is an entertainment for members of a Cambridge college. Its author is unknown, although Moore Smith hazarded an attribution to Thomas Tomkis of Trinity College.[3]

J. W. Clark, *The Riot at the Great Gate of Trinity College, February 1610–11*, Cambridge Antiquarian Society, Octavo Publications, 43 (Cambridge, 1906).

[1] Hilton Kelliher, private communication to the editors.

[2] A sweetball was 'a ball of scented or aromatic substance' (*OED*, adj. C.1). The barber in Middleton's *Anything for a Quiet Life* (published in 1662, but probably written *c.*1621) is called Sweetball.

[3] Moore Smith, p. 8 (work cited on p. 133, n. 4).

Preist the Barbar. Sweet Ball his man. [FOL. 48b]

P. What Sweeteball where are you?

S. Vnder yor nose Sir.

P. Who was heere to speake wth mee since I went?

S. There was three or fower y^{t} brought halfe crownes for you

P. Honest fellowes Ile warrant you halfe crownes for mee?

S. I Sir for you to heale.

P. O knaues then ile warrant you, butt thou wouldst haue said, crackt crownes

S. Why is a crakt crowne, any better then halfe a crowne!

P. Butt before they haue them whole crownes againe, it shall cost them an Angell butt was there no body else heere?

S. Yes, heere was one for you to goe to ye Colledge, to dresse a skollars head they say its broken.

P. I marry S^{r}, att the Colledge, there's a pittifull head, it had need to bee looked after, for it will infect the whole body else, I vnderstand its a ringworme that his head is troubled with.

2 *Sweeteball*] *al* heavily inked 6 *halfe*] *h* written over another letter, possibly *y* or *g* 8 *crackt*] upstroke on *a*; ^{2}c is not normally formed, perhaps begun as *r*

S. S^{r} heere was a gentlewoman too y^{t} said shee was troubled wth ye Collick.

P. A Gentlewoman troubled wth ye Collick is ye wind there ifaith?

S. Shee saide the stone troubled the passage of her water I heere say, shee hath done wth all ye Physitions heere about, and now is come to you.

P. The oyle wthin vpon ye Cubbard head is good for ye Sweelling that a stoane hath made, but whats that to the Collick Are you sure shee's a gentlewoman, why then neatsfoot oyle is very good for her, yett what's that to her making water, Ile tell thee what Sweetball: if shee come againe tell her this receipt from mee. Lett her take a pretty quantity of good roseleaues, and putt them into a still, and keepe a moderate fire vnderneath the still, and ile warrant her my life for hers shee shall make water, and good water too if shee looke too it,: any body else heere?

S. Heere was an other for you to draw a schollars tooth.

P. I'me loath to goe about that busines, lest putting my finger into his mouth, I bee held there tooth and nayle: but I wonder why these schollars haue more teeth drawne then any else, is it because they should haue no more, then to eate there com̄ons withall, for if they haue nothing to bee chauing of theile bee snapping, and bitinge att there fellowes, but is it one of his foreteeth, or his hind, if it bee one of his farthiest, its daungerous.

S. Its a n'ey tooth they say.

P. If it bee a n'ey tooth I shall see itt the better

S. S^{r} they towld mee y^{t} you must sett a ioynt too.

P. What a legg or an arme?

S. I know not butt they said you must sett a boane for somebody.

P. Oh I smell the matter now, I lay my life y^{t} sombodyes nose is putt out of ye ioynt att the Colledge lately, therfore, fetch the oyle y^{t} standes in the caskett, its very soueraigne for y^{t} hurt.

S. Heere was a Gentleman, y^{t} asked if this were Preists shopp.

P. Why you little shaueling you, I hope you could haue told him that it was soe.

17, 18, 22 *Collick*] altered from *Collect*; *i* written over *e*, *t* altered to *k* 22 *Collick Are*] alteration of *t* to *k* in *Collick* obliterates punctuation 27 *it,:*] *sic* 29 *to draw*] *to d* heavily inked 33 *chauing*] *sic* 34 *att*] ^{1}t not fully crossed; *att* in lines 41, 74, and 75 similar 36, 37 *a n'ey*] perhaps *an'ey*

S. So I did, and then hee askt if you were a Surgeon and went his way. Exit.

P. When did you euer know Preistes y^t were not Surgeons, mee thinkes hee
might soone know mee, for I haue done as greate cures as any man, I know
some y^t will take great cures in hand, y^t know not how to performe them, &
haue bin glad themselues to bee cured, by mee, but Sweetball, what are you
gone from the Bason? come quickly or Ile scowre you.

Enter Sweetball wth glasses.

Come y'are a notable sweetball to stay solong, letts see y^t glasse there,
When I was at Sorborne in Fraunce I cured a whole Colledge wth this
oyle. S. Why are Colledges bodyes?

P. I and diseased on's too, a Botts on them, they are much troubled wth ye scuruie
but I hope y^t horses disease hath left them now, I cured ye head there of
a.
a giddinesse in ye braine, as for the body it was much troubled
in ye Reynes, [FOL. 49a]
wch I help too, ye hart of ye Colledge was sound but it was still troubled wth a
bad liuer, the sides had a Conuulsion, as for ye walls they had the falling
sickness, the
Windoes, had a great paine in them, wch I cur'd wth this glasse of oyle of Vitreall

S. But what shall I say to the fellow when hee comes y^t has ye sore throate, y^t hee
can't drinke for't.

P. O bid him take some of ye oyle of Swallowes, and ile warrant him, and to gett
him a good stomack lett him take some of this oyle, I know his ignorance will
think it to bee Sallet oyle, butt hee shall nere know otherwise for mee. Then
for ye Sergeant y^t was hurt bid him looke to his oyle of Mace its very good
for him: And ye Wench y^t was troubled wth ye lasinesse in her bones, lett her
vse this oyle of Broome, & itt cannot choose butt helpe her, as for ye other
Oyle there sett it vpp safe againe: it's very pretious, it has heald I know nott
how many, I knew an hundred at least y^t had ye Pox: Some foureskore a
burninge Feauer. Threescore att least troubled wth an ague. Fortie w^{th}
ye Plauge, And I'me sure twenty wth ye goute: and all these were healed
wth very little, or as a man would say, with none att all of this oyle. Butt

47 *Surgeons*] glossed, with a cross, as *Sr Johnes* in left margin of l. 48 in another (contemporary) hand
51 *Bason*] *o* modified from *e* 54 *Sorborne*] *e* added *Fraunce*] *F* altered from *f* 56 *Botts*] *B* altered from *b* 58 *giddinesse*] *g* written over an erased letter *was*] written over an erased word

I forgett my selfe, they stay for mee by this time att yᵉ Colledge:
Come giue mee
the linnen, the Comb, and glasse presently, for now I must goe to the Colledge
It must bee new trimmd, & bring some Water, I thinke yᵉ Colledge will bee
wash'd very faire and white, but where's yᵉ Comb?

S. There's inough all ready att yᵉ Colledge Sʳ

P. O they be but cox-combs there, why heeres none of yᵉ Raysor neyther

S. Will they vse any att yᵉ Colledge?

P. Vse Raysours att yᵉ Colledge? where else,? why they're verie often shaued there.

S. Will you haue yᵉ Curling iron Mʳ

P. No, they vse to bee cutt very short att yᵉ Colledge in all thinges, they care nott
for any great topps, nor top gallants, yett I must carry yᵉ glasse, for they
loue to looke into euery thing, with out the glasse theyle neuer looke into
themselues, but looke you bringe yᵉ salue after, for yᵉ Colledge must
bee playstred, tis much broken out of late. Exit.

S. Yee had a Bason, yᵉᵗᵗ there's no man washt,
Wee gaue no water least wee should bee dasht:
Yett water for yoʳ hands wee'le nott denie
They beeing wett, wee hope you'le clapp them drye.

82 *else,?*] *sic* *they're*] perhaps *the'y're* 85 *top gallants*] perhaps *topgallants* 89 *Bason*] *o* altered from *a* 91 *Yett*] ¹*t* not crossed; see *att*, l. 34

Item 7 in J.a.2: Fols. 49b–50a *Gowne, Hood, and Capp*

Gowne, *Hood*, *and Capp* is transcribed in hand C′ (see Plates 6 and 7, and above, p. 130). With its references to proctors, freshmen, bachelors, and schools, the play is very much an academic entertainment. During this period, the preoccupation of the authorities with student attire led to the passing of countless (and often disregarded) regulations at Oxford, Cambridge, and the Inns of Court. The play is undoubtedly a Cambridge one, attested (as Hilton Kelliher points out) by the references to a Sophister (l. 63), a Cambridge undergraduate in his third or fourth year, and the reference to playing football 'ath' leayes' (l. 25), or leys, the common land lying at the time behind Pembroke College. As with *Ruff*, *Band*, *and Cuff* and *Preist the Barbar*, *Sweetball his Man*, the play is short, has no female roles, and requires no special machinery for production. Its author is unknown.

Gowne, Hood, Capp. [FOL. 49b]

Enter Gowne, Hood blind folds him.

H. Who am I? G. I know not, oh I smell you now by yor Rose, its shooe.
H. No, gesse againe. G. Oh I feele a hole aboue the shoe its stockinge.
H. No neither, yett againe: who am I?
G. Yo^{r} vile stronge who so ere you bee. Come lett mee loose Garter.
H. On gesse more
G. Sfoot it can't bee Breechees, heele come no more I trow, hee was clapt sufficiently
last night, nay prethee grype mee nott soe hard, I feele you to bee a man of
armes, oh its Dublett.
H. Y'are to wide ith sleeues Gowne for dublett, Yett once more.
G. Ha haue I felt you att last, yett I thinke nott neyther, yes tis, I feele him by
his wooll, tis hatt, come off hatt, of hatt, pheu ist you, what doe you hoodwinke mee.
H. No no, you'le not know your freindes.
G. Nay bee not angry, you know non videmus manticæ qd a vergo est.
H. Who I angry faith Sir no, hood can wincke att all faultes.
G. But Hood where's Capp. all this while y^{t} wee might goe to the scholes
H. Ile warrant you has one odd corner or other to bee in.

Enter Cap.

G. But looke yonder hee comes.
C. Pheu Robin hood fayth how dost thou?, how doe's little Iohn.
H. I mayny will talk of Robin Hood that ne're shott in's bow.
C. And you heere to Gowne? I was all ye markett ouer to looke this long tayld machrell
G. Faith Cap sett all on a merry Pin.
C. I lads you shall find mee a mad cap.

2 *blind folds*] perhaps *blindfolds* *againe*] *in* minim distinction unclear 8 *last*] perhaps *Last*
13 *you'le*] apostrophe very slight 15 *hood*] *h* imperfectly inked 16 *scholes*] *h* written over blotted *s*
17 *bee*] tops of *ee* weakly inked 19 *thou?,*] *sic* 23 *mad cap*] perhaps *madcap*

H. Why where hast bin now Cap, tha'rt so merry?

C. I was ath' leayes att football, and wee had such fine sporte, I playd so well that they all cast there Capps att mee.

G. But I heard say thou wert throwne extreamely, Cap was throwne vpp oftner then any one there, there was nott a meete but Cap was cast vpp, they should haue sent for mee.

C. Alas Gowne what wouldst thou haue done there, they would ne're haue lett thee playd, they cast of gowne before they went to play.

H. Why methinkes Gowne should play well, hee h'as good shankes, would endure a blow well.

C. Hee can neuer throw a man, this is all ye good gowne can doe, hee will hang clogging on a mans back.

H. Heele runne through ye thickest of them, I haue seene gowne brush excellent well.

G. Gowne scornes to budge for ye best of them all.

C. I know not what you scorne but I'me sure I haue seene you taken vpp Gowne as often as any.

G. Like enough for theile play such foule play to gowne, theyle come and take mee vpp behind, when I nere think on't, but if they'd come before mee, I'd nere care.

H. Nay they may doot before you too, for you ly very open before Gowne.

C. Well if hee had bin there now, hee'de a bin glad to a runne, for ye Procters came, and there was the finest sport, one runne this way, an other that way, the Proctour had like to a binne vpon doubletts skirtes, and Breeches had like to a fallen downe for hast, Shoes tooke his heeles, and runne away, as fast as hee could, & hatt lost his band for hast: and nobody durst stand ye Proctour but Capp.

H. Come capp wilt goe to the Scholes, thou hast nott bin there a long time as I haue seene

C. Nor you neyther if the truth were well knowne.

G. Yes hee kept acts the last weddensday. C. where ith Philosophy scholes?

H. Hood scornes to keepe actes wth freshmen: or wth any other but Bachelours

25 *such*] *c* written over blotted *f* 27 *thou*] tail of *u* continues for another minim; perhaps *thow*
32 *hee h'as*] *sic* 35 *clogging*] *l* is written over another letter 51 *weddensday*] perhaps *Weddensday*
where] perhaps *Where* *scholes*] *h* altered from *g*

G. No ile warrant you, you thinke hee should doe as you doe: you nere doe any thing but ith night-Capp.

C. In troth I know not whether you do any thing ith' day or noe, butt itt might bee.
bee ith night-gowne too for any thing I know, and soe wee might both keepe our [Fol. 50a]
actes in tenebris.

H. A way foole a way, yor coxcomb shewes you to bee but an Asse.

C. An Asse fryer bald pate? G. I and a blockhead Capp.

C. Heere's fine stuffe gowne, Ile tell thee what Gowne, Cap scornes but to haue a better head then any of you both.

H. I thinke you would haue sayd a greater, for Im'e sure you nere durst doe any thinge ith scholes, why whilst you were Sophister you were suspended Capp euery day for one thing or other.

G. Nay and you bee soe round cap wee can take you downe. as for yor schollarshippe I thinke you were nere out of Aristotles Ethickes yett Capp.

C. why thats more then euer you learnd yett, you nere knew what good manners meant nay Gowne Ile warrant you yor facing of itt shall nere carry it away soe.

H. I prethee Cap tell mee the reason you satt not ith schooles in Lent.

G. what not a worde, Oh you're a considering Capp, are you.

C. Fayth Ile tell thee what Gowne, methinkes euery one should cutt his coate according to his cloth, what if't cost capp a crowne to buy out his settinge, did not you do so too Gowne?

G. No Sir I satt all ye while.

C. Nay itts noe matter, if you had hung ith' meane time, but indeed I think you did but
sett you went vpon none.

H. Well Cap you need nere disprayse Gowne, for I can tell you hee hath bin putt to ye proofe.
There are feaw M^{rs} of Arts ith' towne but will make much of gowne.

C. Alas Hood you had more neede speake for yo^{r}selfe for gowne is well backt wthout you.

54, 56 *night-*] *ni* minim distinction unclear 58 *A way . . . a way*] perhaps *Away . . . a way*
67 *why*] perhaps *Why* *nere*] perhaps *ne're* 70 *what*] perhaps *What* 74 *time*] *im* minim distinction unclear *indeed*] *in* minim distinction unclear 78 *yo^{r}selfe*] perhaps *yor selfe*

H. Why I hope you make no doubt of my learning?
C. How canst thou bee a schollar when thou art a riding-hood so often: thou sat'st in Lent only for fashion sake.
H. Its well knowne what I am.
G. I ile beare Hood witnesse hee went ouer all ye Bachelours ith' scholes, before hee had donne.
C. Nay nere speake you so much for Hood, hee'le bee vpon yor back too, Gowne ere it bee long. Come I know him of old. H. And what do you know him for?
C. Marry for a Notorious fox catcher. G. Why I think Hood nere goes toth' Tauerne.
C. No, but I'me sure hee goes toth' Alehouse. Hood is altogether for Lambswool⟨e⟩
H. Speak no more Cap then you can proue.
C. No more I do'nt. for I my selfe haue oftentimes seene you a casting-Hood.
G. Why Cap you need not speake so much against hood for thatt, I haue knowne you bin a foodling-Cap too.
C. Fayth to speake truth of thee gowne, I must needs say I neuer see thee foxt, but I mett you Gowne, and Staffe wth an old man once, & I faith Staffe was well whittled, & you lind: well you had best leaue itt, and say I tell you in freindly manner of it, take heed you'le gett a habitt ere it bee long, and then youle haue yor Hood puld ouer yor Eares, and then wee shall haue a mourning Hood.
G. I hope this geare will cotten better with Hood then soe. Exit Hood.
C. But where is hee, what no sooner talking of being drunck, but hee's gonne!
G. Why hee's gonne tooth scholes before, doe not you know y^{t} hees to reply to day, and soe hees affrayd to stay to long. C. Doe you gowne answere, then indeed Hood will come vpon you.
G. Yes. C. Then I preethee goe follow him, and Ile follow you, presently. Exit Gowne.
this tis to bee great ith Proctours bookes as Cap is hee has made mee Moderator, and now I haue a

80 ^{3}*thou*] *o* written over an erased *y*; ? originally *thy* 86 *Notorious*] *N* inked over another letter, perhaps *n* *fox catcher*] perhaps *foxcatcher* 87 hole in Fol. 50 may have eliminated end punctuation 89 *do'nt*] *sic* 90 *against*] *n* written over another letter, perhaps *a* 99 *tooth*] perhaps *too'th*

good time to bee quitt wth Gowne and Hood both: If Gowne bee tedious in's Position, ile cutt of Gowne, ile nott haue him to long. As for Hood ile take him of presently if hee Dispute not well: but I know they stay for mee.

Our Author bids mee say for's Gowne, and Hood,
It is the Taylours fault, they are not good
But howso'ere for feare of wors mishapp
Hee lowly craues a Pardon wth his Capp

103 *time*] *im* minim distinction unclear